Permaculture Design Companion

A Practical Workbook
For Integrating People & Places

Jasmine Dale

Permanent Publications

Published by
Permanent Publications
Hyden House Ltd
The Sustainability Centre
East Meon
Hampshire GU32 1HR
United Kingdom
Tel: 01730 823 311
Fax: 01730 823 322
International code: +44 (0)
Email: enquiries@permaculture.co.uk
Web: www.permanentpublications.co.uk

Distributed in the North America by

Chelsea Green Publishing Company, PO Box 428, White River Junction, VT 05001
www.chelseagreen.com

MIX
Paper from responsible sources
FSC www.fsc.org FSC® C007785

Printed in the UK by Bell & Bain, Thornliebank, Glasgow

All paper from FSC certified mixed sources

The Forest Stewardship Council (FSC) is a non-profit international organisation established to promote the responsible management of the world's forests. Products carrying the FSC label are independently certified to assure consumers that they come from forests that are managed to meet the social, economic and ecological needs of present and future generations.

British Library Cataloguing-in-Publication Data
A catalogue record for this book is available from the British Library

ISBN 978 1 85623 329 3

"step by step,
breath by breath,
hand in hand
we weave our web"

Starhawk

o0o

Huge gratitude to Simon Dale for the design and layout, bringing this book to life,
and to Jude Dunn for editing support

CONTENTS

How to use this book 4

PART 1: Awakening Our Inner Design Beast 7

What is Permaculture and Why is it Useful? 8
- History & Origins of Permaculture 9
- Reflections on Wildness 10

Observation 14

Patterns 19
- Branching 22
- Sphere 23
- Spiral 24
- Radial 25
- Web 26
- Honeycomb 27
- Torus 28
- Random Scatter 29
- Applying a Pattern to your Design 31

Permaculture Principles 32
- Observing the Principles in Nature 34
- Applying the Principles in Design 40

PART 2: Seeing Things Anew 43

Design Processes 44
- SADIMET Design Process 46
- Exploring Complexity 48
- CEAP 50
- Appreciative Inquiry 50
- OBREDIMET 51

Vision 52
- Benefits of a Vision Board 54
- Making a Vision Board 56
- Vision Statement 58

Landscape Survey 60
- Other Landscape Survey Resources 65
- Listening to the Landscape 66

Mapping 68
- Mapping Techniques 70
- Learning the Slopes 75
- Additional Resources 77
- Base Map Checklist 78

People Survey 80
- Elements Survey 81
  - ~ EARTH: Skills & Resources 82
  - ~ AIR: Thoughts & Ideas 83
  - ~ FIRE: Passion & Enthusiasm 84
  - ~ WATER: Transitions & Flow 85
  - ~ SPIRIT: Unique Qualities 86
- Traditional Client Survey 88

PART 3: Weaving the Web 91

Introduction to Analysis 92

Key Planning Tools 93

- Sector Analysis 93
  - ~ Soil Sector: Earth 94
  - ~ Wind Sector: Air 98
  - ~ Water Sector 102
  - ~ Sun and Shade Sector: Fire 106
  - ~ Wild Sector 110
  - ~ View Sector 114
- Slope, Elevation & Aspect 120
- Zone 124
- Network Analysis 127

Input <-> Output Linkages 129

Elements & Functions 132

Synthesis 138

PART 4: Going Forth 141

The Design 142

- Template Design Tool 142
- Creating Guilds 144
- Applying Permaculture Principles 152
- Concept Design Map 160
- Detail Design Map 161

Implementation 162

- What, Why, When, How, Where & Who 163
- Maintenance: Troubleshooting 164
- Action Learning Plan 166

Evaluate and Tweak 168

An Ethical Foundation 170

Next Achievable Steps 172

Additional Resources 175

- Landscape Survey 176
- Managing Complexity 178
- Brief Comparison of Design Frameworks 180
- Summary of Analysis and Design Tools and Methods 181
- Repair Plants: Dynamic Accumulators 182
- Permaculture in Practice 184
- Deep Time 188

How to use this book

This is a practical workbook to mentor you through a personal project, following all the steps required to create a thorough and effective permaculture design. It is a helping hand to complement the excellent courses and books already out there, to convert enthusiasm into practice. To deepen and create connections between people and places.

This book works whether you've heard of permaculture or not; whether you wish to make a difference to big issues, such as climate change, or are more concerned with making your own project enjoyable and in harmony with the world around you. Maybe you require a taskmaster to get something done and a route to follow, or you are interested in design and natural systems. Whatever brought you here, this is a tried and tested process to creating a coherent, relevant and engaging design.

By working through the book and completing the exercises, you will understand:

WHAT each step means

WHY it is relevant

HOW to apply the idea or tool

WHERE everything is best placed

WHEN and in which order to action the design

WHO is involved.

Choose a project you wish to design. The exercises can be applied at any scale from a small patio, to a garden, smallholding or a very large site, such as a woodland, farm or even a bioregion.

You can start from wherever the site is at now, whether a blank canvas; midway through a design or project already underway; or when picking up the reins on a site where activity is established.

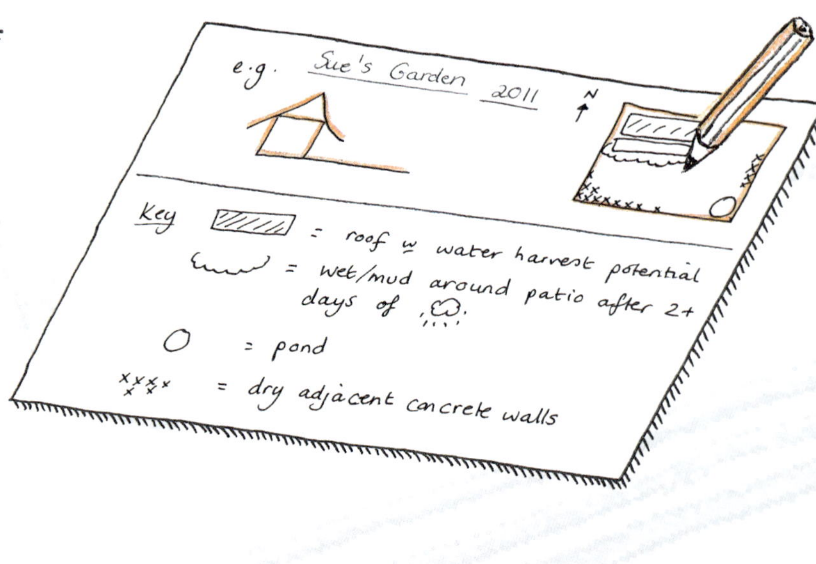

You are the designer, however you are not an island and the exercises will incorporate other people involved in your project. This book can also be used by a group to plan and action a design together.

Begin to dig deeper for your motivations in designing for your chosen place. The more honest we are here will inform clear and useful outcomes. Harmonising big concerns with our personal and immediate desires creates elegant and exciting designs.

Work through the exercises sequentially. Lots of your findings and ideas will be recorded in this book, though not all, so an A4 book for supplementary analysis and research is useful. Some exercises, such as mapping, will be on big bits of paper– so a folder may be a good idea.

Although this book is linear, design, like nature is cyclical and has complex interactions. Observation, analysis, design and feedback sometimes happen simultaneously. The exercises flag up the need to revisit previous steps when and where is most appropriate.

PART 1:
Awakening Our Inner Design Beast

What is Permaculture and Why is it Useful?

Gardening and home-making are the oldest and most widespread crafts on earth. For more than 9000 years techniques have been developed that have kept the human species in balance with the abundance of the land. Since the first upright ape, Lucy, 3million years ago, we have consciously stewarded our environment to meet our basic needs of shelter, water, food and energy. Now our species is at a crossroads, a Great Turning point in the history of humanity: the life support systems we depend on are threatened with collapse by the consequences of our own activity.

Permaculture offers us a creative route out of this situation; a systematic map to go beyond survival and create 'thrival' on this earth. A permaculture practice provides thrival by harnessing our intellectual and imaginative potential, whilst drawing on the principles of our ancestors and wild ecosystems. We do this is in combination with awareness of the wider, context we live in Now and of our own personal, specific circumstances.

There are many definitions of permaculture; here is one I use:

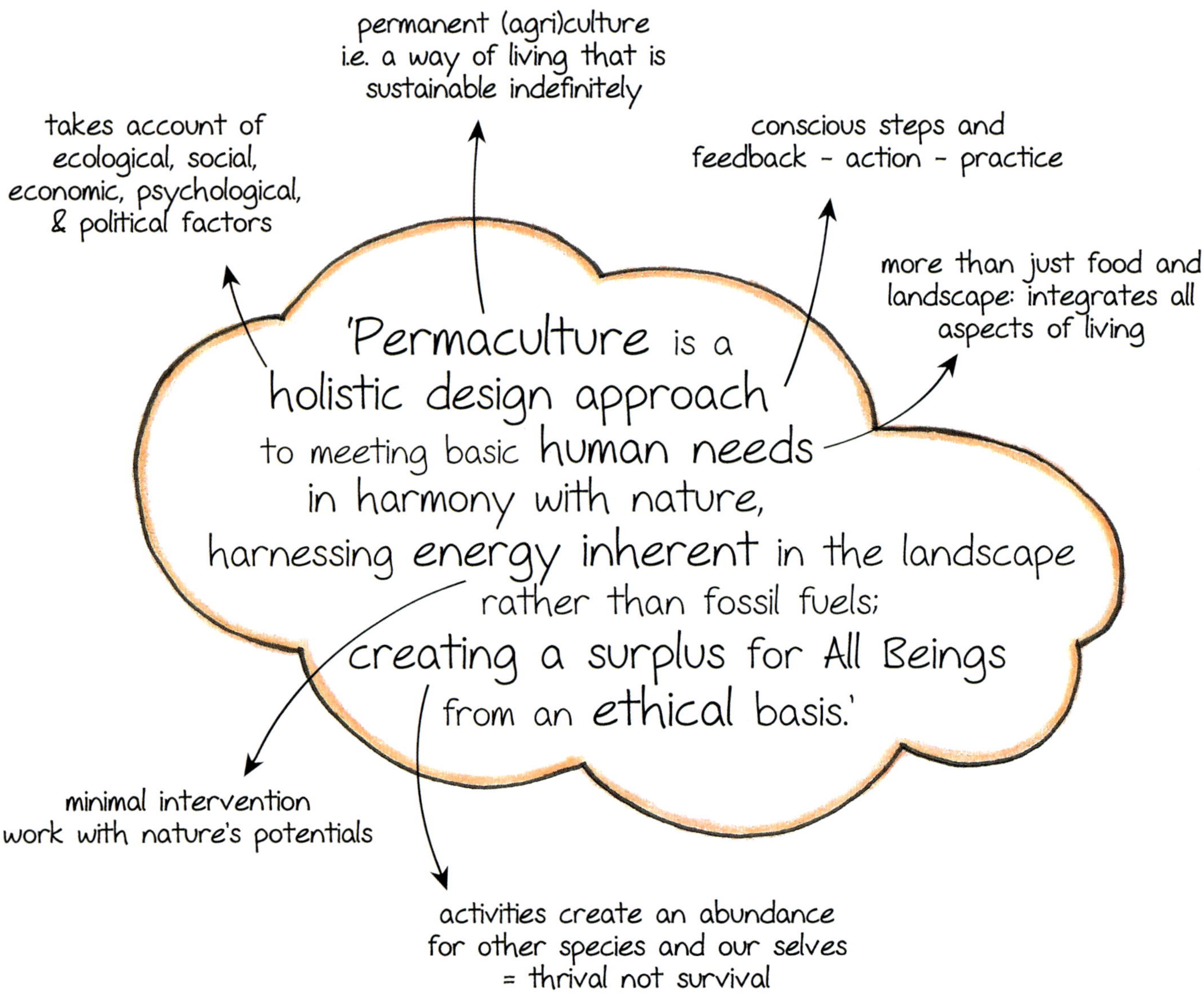

History & Origins of Permaculture

In the early 1970s the myths of economic growth, driven by fossil fuels, received a wake up call. Oil prices quadrupled in one year, a 4 day week for industry was imposed and domestic electricity was rationed in Britain. The Club of Rome, a distinguished bunch of international economists, published the report 'Limits to Growth', which called into question fundamental assumptions of an oil based economy and the promise of perpetual growth.

Ecological collapses, species extinction and pollution were more and more evident and blatantly caused by our industries and land use. The great social changes already underway for civil rights, peace and disarmament were joined by a growing 'green' movement and at a governmental level Sustainable Development was born.

Meanwhile, in the Australian bush, Bill Mollison was carefully observing the natural systems around him. Bill had worked on the vast felling of forestry plantations as a young man and saw how bereft of life they were and how much energy they required to sustain and harvest. In contrast, Bill observed the wild rainforest, with its complex web of interactions, creating zero waste, requiring zero oil inputs and yet producing diverse habitats and thrival for a wealth of species.

Nearby, David Holmgren, a young student of Environmental Design, was exploring radical ways to solve problems in this field. Together with Bill, they created a body of work that modelled wild abundant systems to meet human needs without the use of fossil fuels. Permaculture was born in the form of the book 'Permaculture One' in 1978.

Grounded in the earth sciences and General Systems Theory, they applied their method on a landscape scale, identifying solutions for the vast expanse of the Australian continent, designing for minimum input - effort and maximum output - gain. However, the concepts of permaculture are applicable on any scale, as the founders intended. Today many of the best examples of permaculture design are found in city gardens and small homes, limited space being a good impetus to careful and intelligent design.

PARAMAETHU

The welsh word for permaculture is derived from amaeth (agriculture or cultivator) and parahaol (permanent). These terms link beyond farming to culture in the verb maethu, to nourish or foster, and the colloquial para, to continue or last. Adopting the word in its verb form, rather than as a noun, adds to the feeling of an active process as paramaethu directly translates as 'permaculturing'.

www.konsk.co.uk

Reflections on Wildness

The core components of permaculture are outlined below, derived from Bill and David's keen observation of the work, or processes, of the forest and reflections on the essence of wildness. They brought mindfulness of how humans work and play to the heart of how ecosystems thrive. They envisaged immediate solutions to the tangled briar of modern problems and of reducing 'work as toil'. With this in mind, permaculture celebrates human well-being and delight for life as part of our wild heritage; our birth right. Crucially, this is achieved without compromising the well-being of other species or the landscapes and natural systems that sustain us all.

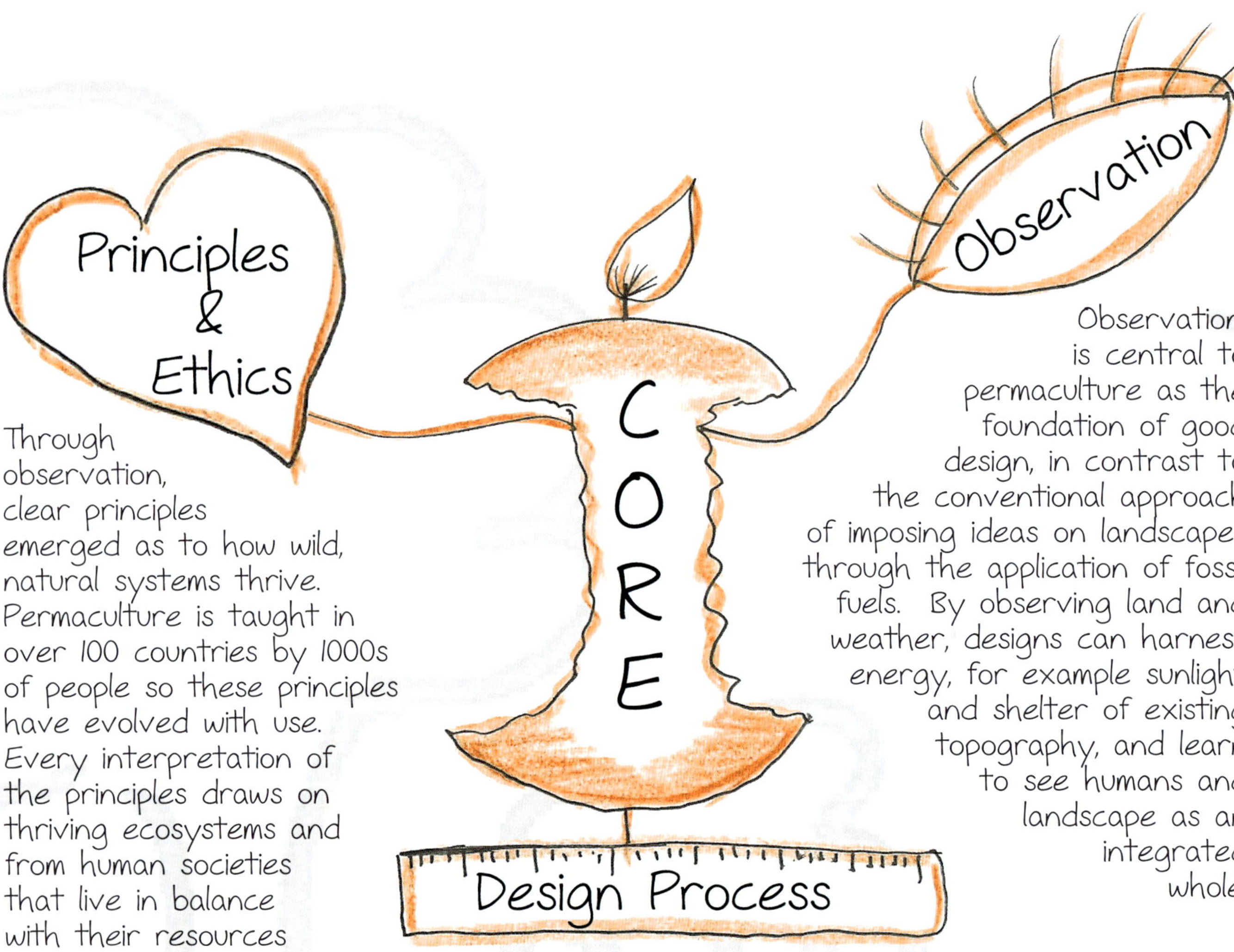

Through observation, clear principles emerged as to how wild, natural systems thrive. Permaculture is taught in over 100 countries by 1000s of people so these principles have evolved with use. Every interpretation of the principles draws on thriving ecosystems and from human societies that live in balance with their resources (mainly indigenous people).

Observation, is central to permaculture as the foundation of good design, in contrast to the conventional approach of imposing ideas on landscapes through the application of fossil fuels. By observing land and weather, designs can harness energy, for example sunlight and shelter of existing topography, and learn to see humans and landscape as an integrated whole.

This sets permaculture apart from other approaches to sustainable living, such as organic gardening, by defining a conscious design process – at its core the design process is an observation/survey phase; followed by analysis and design; leading to evaluation and feedback into further activity.

Throughout antiquity the majority of people lived and worked in balance with the resources available in their local landscape, akin to the principles of permaculture. There was little other choice. Over centuries we drifted; slavery, colonialism and later fossil fuels engendered a laziness that unravelled the common sense pervading the design of our early settlements.

'APPROPRIATE LOCATION' is a key idea our ancestors understood well.

Everywhere, the energy potential in the landscape was harnessed for water wheels and mills, transporting goods by river; farms, villages and towns sprouted where they could thrive, just like trees and animals do. Homes and gardens grew where fresh water, fuel and shelter would provide resilience and livelihood, safe from floods and high winds. People lived in beautiful places that were good for the soul.

All activities and infrastructure were part of a WEB OF BENEFICIAL RELATIONSHIPS supporting the functions of each other, utilising any 'waste' back into the system as a useful resource. Buildings, work, energy, food, crafts and industry were diverse and overlapping spheres of life. Woodlands, farms and gardens were polycultures supplying multiple needs, such as pollination, fertility, food and medicine.

Produce and crafts evolved with a LOCAL relevance, specific to the available materials and culture in a region. Architecture reflected the resources and constraints of the landscape and skills were passed down the generations.

Is it possible to imagine such a world in medieval times but not today? Even if you can picture a revival of the best of the past, there remains the question, how fair was society for our ancestors.? You might point out that although we no longer drown witches, prejudice, suffering and domination persist through the centuries. Many people feel resigned to the feeling that humans do not appear to be evolving, despite technological advance. Plus, it is reasonable to think new challenges abound like climate change and global instability that render a return to the balance of our ancestors' world to be impossible. Plus, I imagine you don't wish to return to a world of turnip pie, open sewers in the streets of London, the black death and no internet.

That is understandable. A shift to a permaculture world is going to reflect where humanity is now, rather than a wholesale return to the past. We have more awareness than ever before. There is more social diversity and free thinking in society than our 12th century villagers could comprehend. There's plenty of scope for ingenuity and an unprecedented quantity of spare stuff to counter balance the challenges we face today. The legacy of mass manufacturing is not all doom and gloom, think of all those mountains of double glazing and stainless steel sinks waiting to be re-used! In agriculture, education, communities, business, technology, art, gardening, and architecture, progressive change is occurring. Between individuals, families, groups, religions, cultures, nations and continents we are having to learn to cultivate trust and peace. Inspiring initiatives pop up everywhere daily.

HERE IS A PEEK INTO A POSSIBLE PERMACULTURE FUTURE...

At the beginning of the 21st century, the last old timers tales were told and the last wild, secret places were subdued and the remaining old threshers, looms and tools were carefully placed in museums. Meanwhile, a new seed began to germinate in the hearts and minds of millions of people across the world.

Within a few years we learnt a house could be warmer by facing the sun and sheltered from the wind by trees, even in the city. Vast woodlands were grown all over the land, joined by a network of wildlife corridors through our towns and cities. Suburban gardens left narrow edges along their fences and walls, unmown for creatures to cross the land unseen, to find food and romance under the cloak of night. Birds, insects, small mammals and wild plants thrived, reversing the tide of species extinction. Quickly the woods calmed extremes of heat, wind and heavy rain; keeping the precious water in the soils, off the roads and out of people's houses; rebalancing the greenhouse gases in the air. The star drenched night sky returned as streetlamps cast their gaze downwards and buildings switched off their lights for a rest.

Vacant buildings became centres for local commerce and skill sharing, people of all generations and walks of life meet there to buy their food, swap their produce and have a good time. Useful machines, such as apple presses and 3D printers, are cheap to use and people take turns to run them; they bring in their latest ideas and share the knowledge with the community. People who are rubbish with new gadgets are helped by those that understand them. Trading is flexible and simple, trust has replaced paperwork to a large degree. Work and play are not so far apart in how they are perceived. School has meaning, work has meaning, we enjoy it.

Nothing much is wasted any more; whether it's heat recycled from buildings and industry or sewage and manures providing fertility for crops and energy; reed beds creating valuable habitat. Toilets are flushed with rainwater, autumn leaves are composted close to where they fell and later become living soil at the nearest garden or field. Dogs receive bones direct from butchers. Office workers are kept warm by the motions of those attending the gym, they grow chillies and tomatoes in their break times to replenish the air, for fun and to contribute to the local economy and food supply.

Public places are largely pedestrianised, full of fruit trees benefiting from the concrete microclimate and plenty of passers by to eat the harvests. Rainfall soaks in to channels adjacent the pavements and fruit trees, forgoing its pointless journey to the sea in favour of replenishing groundwater aquifers for drinking and irrigation. Local veg producers are subsidised on every high street as the global food economy rebalances and supermarkets adapt to more regional supply.

For future generations born into this world it is all common sense. Conserving resources, joined up thinking and redistributing surpluses is as natural and obvious to them as the existence and essentiality of mobile phones are to our children.

Permaculture is flexible, a lens through which we can view our world. It is an attitude and a framework, a pattern language for ushering in an Ecological Age. Like nature, permaculture is about creativity and harmony, so if you find yourself stuck with your design – chillax! As one of my teacher's said, "permaculture is a game we can play. We can say to ourselves, today I'm going to work this out with the permaculture game."

So, let the games begin!

Complete the following statements

Where? I/we are designing:

Why? I/ we wish to do this:

And to get us started with observation, evaluation and feedback:

The main idea that I remember from this chapter is:

The feelings, thoughts or reaction I have are:

Observation

Observation of your site is the process of building a relationship with that place, getting to know all its quirks, moods and characteristics. Over time this understanding will deepen, so don't feel that everything must be mapped and named at the outset. In other words, allow for your observations to grow, especially as things are not always as they seem at first sight. For example, a dry patch of ground can be waterlogged at other times of year; or a bare, scruffy corner may sprout glorious wild roses and useful plants come spring.

You will need a simple notebook to record your observations over time, in addition to the exercises in this chapter. Something that's easy to have in your pocket whenever you are at your site. It's important to note the date and weather at the time of your observations too.

Nothing else substitutes for paying attention to what's actually on the ground, whether you are designing houses, robots, nursing the elderly or painting fences. All great gardeners are keen observers, hence the saying 'the best fertiliser is the gardener's shadow.' Whatever the task, mindful attention gets optimum results. In an ideal permaculture universe, a site is observed for a year before intervening in any significant way or before making design decisions, to enable your place to be experienced in all its seasons.

If a year of observation is not feasible, we can at least design and implement a project in stages, with small and slow interventions, allowing new observations and information to be incorporated. Especially as once a project is underway factors are bound to arise in response to any activity.

What is the alternative to observation, except for imposing fixed notions onto unknown territory? To work with the landscape and nature's cycles we are going to adopt a formal decree:

Declaration of Observation

name of site ↓

I/We . . . . hereby vow to suspend any assumptions about . . . .

I/We vow to be humble to the complexities and interactions, both visible and unseen at this place.

I/We acknowledge we have ideas and plans for this place, which may get us very excited and prone to action.

I/We declare to undertake a commitment to observation, formally and informally, before I /we make a design and implement it.

Signed by: . . . . Date: . . . .

Having led observation exercises with hundreds of people, I have learnt how frequently humans jump to conclusions and how common it is to come at a place with fixed ideas of what is suitable there. It is a strong temptation to resist. So, to help us return to being Observers before committing to a plan, and to take the next step of becoming a permaculture designer, as your Companion on this journey I suggest the following:

Park all your ideas for your place or project here for now. Be truly open to what the land at your site will reveal and let that guide your design.

scribble them all down, don't be shy!

There are many gadgets and data sources which give us information about a landscape or place, expensive laser levels and complex soil analyses for example – all of which can be very useful. However, there is an incredibly sensitive, highly calibrated instrument available, evolved over millennia to sense and feedback environmental conditions. Everyone has access to this instrument, at all times and with no financial costs or strange buttons to master. Our own Body.

Developing our sensory awareness and becoming conscious of subtleties at our place, brings us back into alignment with our indigenous ancestry and less likely to impose 'top down' ideas from our brains onto the land. As my dear Professor Pocock once explained:

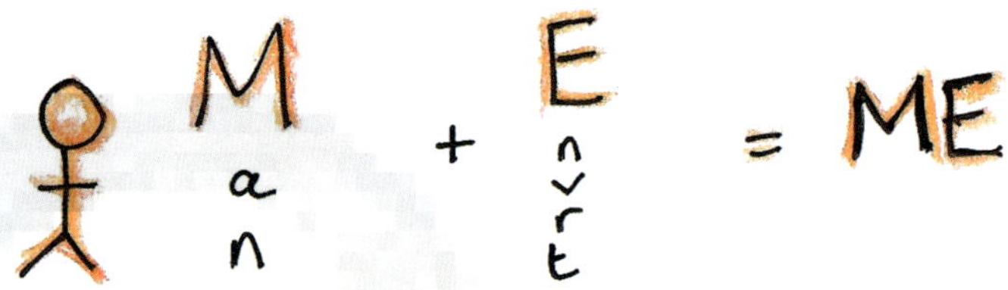

So, as we attune to our senses, an integrated ME arises, that is interdependent with the landscape and thus in harmony with it. Objectivity, observable reality, becomes truly apparent from realising our subjective experience.

Our bodies observe, act and feedback unconsciously all the time, keeping our organs going, blood circulating, enzymes releasing and so on. Even our minds are expert observers; parents, pet owners, good speakers and teachers, cooks, musicians, artists, cleaners, builders, managers, meditators, martial artists, sports people and more rely on mindful observation if they play their role well.

I am a good, competent observer when I am ...

name some activities which you already apply good observation to

There is a wealth of sensory and nature awareness exercises out there, particularly drawing on the Wilderness movement and Native American Indian practices. Here is a simple exercise you can try out RIGHT NOW where you are sitting. All you need to do is get comfortable and be undisturbed for a few minutes:

Feel the ground beneath your feet. Notice its solidity, bringing your attention to any textures, lumps or smoothness.

Extend your minds eye down through the floor or ground as if you have roots, imagining the rocks beneath the soil wherever you are. Imagine if you can beneath the earth's crust right to the molten rock and liquid core of the planet.

Bringing your attention back through your feet to your whole body, notice the surface of your skin. Feel the weight of your clothes and the sensory quality of their texture. Feel the air on your face and skin, its temperature, its movement and origin or direction of flow.

What can you taste? Are there smells you can detect? Of yourself, others, the place around you?

Finally, what can you hear. Take a minute with your eyes closed to really hone into the soundscape around you. Start with the sound of your own breath, the crinkle of your clothes and gradually move your attention outward. Take in the subtlest of sounds around you, extending your hearing to the far distance.

Bring yourself back to your feet and the ground and gently waggle your fingers and toes. Have a moment before you start thinking, moving, talking and so on and enjoy your heightened senses with a deep breath.

Doing this sensory exercise, my impressions were:

Coming present to our senses, induces our brainwaves to a state of 'relaxed alertness' - it is this quality we wish to cultivate as good observers and permaculture designers. Vegetarians, meditators, native peoples and mountain dwellers all tend toward this brainwave. We can induce such a state fairly instantly through using our peripheral vision; this means letting our eyes soften and widening our gaze to encompass 180°. We can find the edge of this field of vision by stretching our arms out perpendicular to our bodies and waggling our fingers until their movement is just visible, on the periphery of our sight.

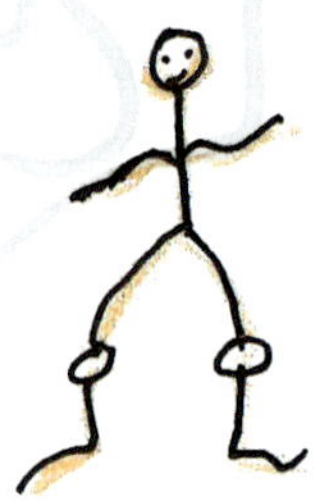

Soft knees, always a bonus

looking at a view upside down also yields an interesting perspective

Noticing adverse reactions or disturbances, is a key indicator to be embraced, rather than switching off and going back inside e.g. freezing winds = lots of info and priorities for design

The next time you are at your Place, give at least 20 minutes to observing your senses (either sitting, standing or walking), in combination with activating your peripheral vision. You can also use this time to observe through the four directions, moving your attention between the horizon, middle distance and close up, to enhance sensory acuity.

The next step on our designer's journey is to pick a Sitting Spot, where you commit to regular, observation practice. When you've found it, give it a name and write it here.

I promise to regularly tune into my senses throughout this design process and keep a record of these observations in a notebook, however mundane they seem.

Signed Date: . . . .

Shamans know a thing or two about awareness, landscape, observation and 'reading the signs'. It is a strong tradition to develop this art by choosing a Sitting Spot for tuning in and observing. You'll know the place when you find it, because it will feel just right.

Patterns

This chapter will look at 8 archetypal patterns that underpin every natural system on earth, including within our bodies. Like the patterns of our minds and language, they take a form to fulfil a function. For example, a river system creates a branching pattern as a surplus of water collects under the force of gravity. Tree roots create the same pattern to collect nutrients and water from the soil. In the branches of the tree, or the delta of the river, the same pattern serves the function of dispersal, the reverse of collection, dispersing the water and nutrients to the places where they are needed. These forms are dynamic and are perhaps better understood as processes rather than the relatively static notion of a pattern.

A pattern's form in a sense shows us the boundary in a system's process, the containment of energy in a certain state or an aspect of a more complex function. When the river flow meets new conditions, such as the sea, it becomes part of another set of forces: tide, moon, wind, global thermal currents, reefs etc. The water adopts a new pattern of activity and interacts with other patterns at work until it reaches a new edge.

Pattern recognition is our ancient birthright, enabling our evolution and thrival on this earth. The reverence and attention to pattern is visible in all cultures from the dawn of human time. The commonality of spirals, shared across the globe, is an easily recognisable pattern, as are other symbols evident in prehistoric cave art to complex Buddhist mandalas.

Modern life, the pace and the strain, the toxicity, noise and digitalisation is often credited with destroying our intuitive pattern recognition faculty. I feel more optimistic than that. We have enjoyed this capacity for millions of years and industrialisation, a mere blink of the eye in evolutionary terms, has little power in the face of the ancient, innate marriage of pattern, process and person.

The challenge as permaculture designers in the 21st century is to reclaim our pattern knowing. The whole body of work of permaculture mentors us back to our intuition and natural expertise as designers and readers of landscape and materials. We can accelerate the recovery of this skill by becoming familiar with nature's patterns, their function and consciously applying them to our designs.

Have a minute to think about what is the predominant pattern in your life right now. It might be changing nappies, work schedules, a seasonal routine, the grain of wood you are whittling, a musical piece you are learning, a recurrent theme on your mind....

My own current dominant pattern is:

Sketch and/or scribble the feeling of it here

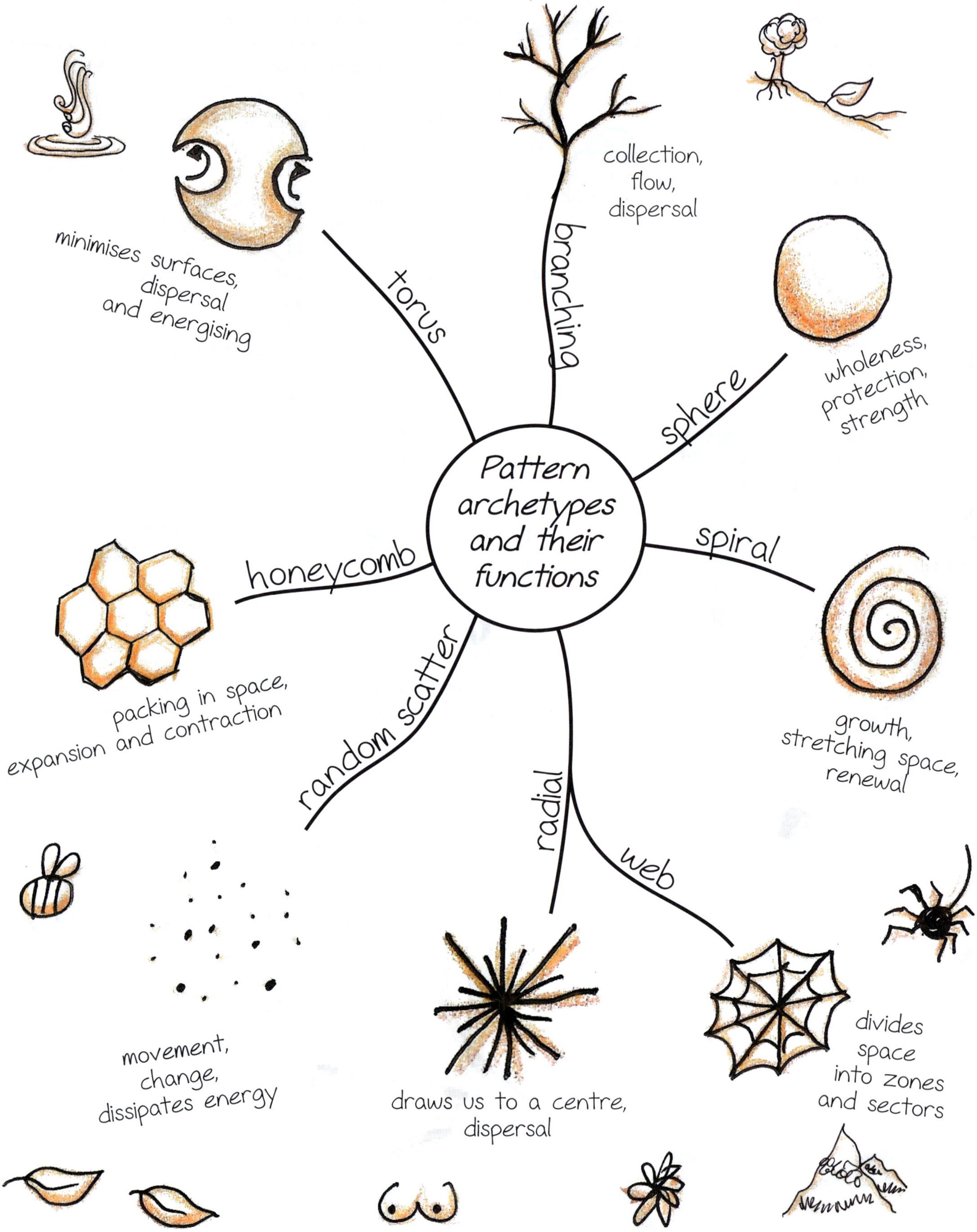
Pattern archetypes and their functions
torus
minimises surfaces, dispersal and energising
branching
collection, flow, dispersal
sphere
wholeness, protection, strength
spiral
growth, stretching space, renewal
web
divides space into zones and sectors
radial
draws us to a centre, dispersal
random scatter
movement, change, dissipates energy
honeycomb
packing in space, expansion and contraction

There are immense bodies of work on patterns if you are keen to develop your Inner Professor and intellectual understanding of this topic. Bill Mollison's 'A Designer's Manual'; any book by Christopher Alexander and David Holmgren's 'RetroSuburbia" are serious tomes, full of fascinating examples of how patterns pervade landscapes, society and even galaxies. Carl Jung and his successor, Joseph Campbell, illuminated patterns as fundamental to the human psyche. Campbell offers an encyclopaedic insight as to the universality of patterns across all tribes and cultures in 'The Power of Myth' and 'The Hero with a Thousand Faces'. Finally, Starhawk's 'Earthpath' has a grounded approach through mind/body exercises to enhance observation of patterns and processes in nature, ourselves and society.

Now observe each pattern to create a more in depth relationship with it in order to understand its function; then we'll naturally begin to apply it in our work.

Top tips for getting the most from this exercise:

1. Devote at least 20 minutes per pattern. Use your own site where possible.
2. Go outside, drop into your observation state through coming present to your senses.
3. Sketch with attention and relaxed alertness 1 or 2 examples of each pattern in the space provided.
4. Give yourself permission for the sketch to come out as it will so you avoid any hang ups about art, drawing and your own skill.
5. Get a botanical lens so you can get a really close up view. An empty loo roll tube is an excellent "nature's eye" substitute for a lens. Either bring focus to the exercise.

OK, let's go do it!

If it's really too cold outside, collect up pattern samples and do it at home

Branching

a 30° branch facilitates movement; intimately linked to the physics of water and flow. From tree canopy, to our veins and lungs, from river valleys to leaves and lightening. Branching is all around us collecting and dispersing energy.

Sphere

minimising surface area to protect from external forces and events, lending strength and a safe environment. From eggs, wombs and cells, to seeds and fruit, from our eyeballs to planets, spherical forms abound.

Spiral

in the growth of tree trunks, snail shells and sea shells; in clouds, tumbling waterfalls and nests. Spirals are a function of strength, renewal and growth; stretching space to enable a lot of activity in one place.

Radial

whether a woman's nipple to a feeding baby, or the nectar in a flower to a feeding bee, a radial pattern is an invitation to come close and be drawn to a centre in order to disperse the treasure. Seed heads, like dandelions, use the radial pattern to disperse their goods across vast distances.

Web

divides space according to activity and conditions. A spider's web is highly efficient for traversing space and communication of activity over an area. The zones across a mountain, or any landscape, differentiated by climate, moisture and rock form a web of flora and fauna innately suited to their location.

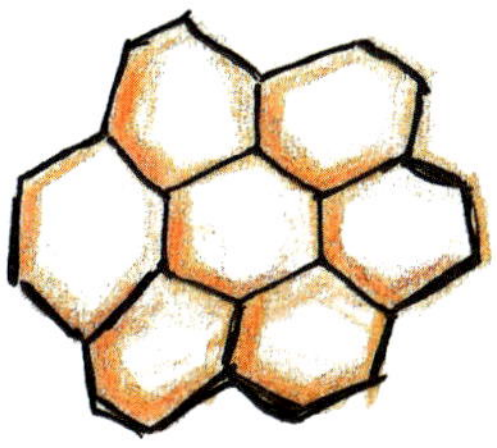

Honeycomb

this 'packing and cracking' formation is visible on bark, skin, land that's been soaked and dried out; a pattern of expansion and contraction and a feature of drainage. It's also the sphere pattern, squashed to provide spatial efficiency and maximum protection, as in a bee hive or eggs in ovaries.

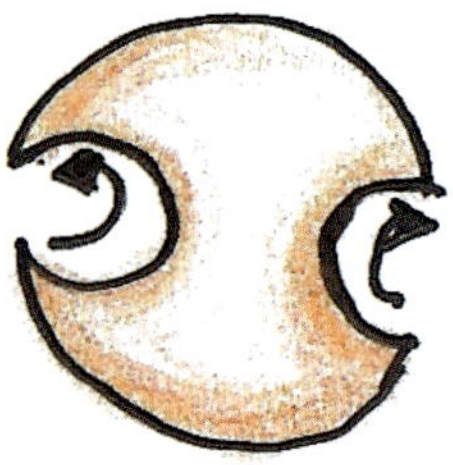

Torus

profound and perhaps least well known, the waveform of torus is all around us. From the transport of water and minerals in people and plants up to the scale of clouds, waterfalls and black holes. Magnetic fields, DNA, fast moving water in streams and big windy clouds all have the torus pattern whose function is to simultaneously disperse and create energy by minimising surface area and generating flow.

(black holes and DNA being somewhat elusive, permission to observe through youtube and / or water descending down the plug hole)

Random Scatter

the swirl of autumn leaves, the eruptions of a volcano and the markings of a leopard; microscopic spores unfurling from a fern or drops of water refracting through sunlight, to the prosaic scatter of sheep poo on a welsh hillside. This pattern's function is to disperse energy, indicating movement and change in a system as matter and energy transition to another state or process.

PATTERN	FUNCTION	DESIGN APP EXAMPLES	IDEAS WHERE I MIGHT APPLY THIS PATTERN IN MY LIFE OR THIS DESIGN
branching 	collection, flow, dispersal	paths, revision mind map	
sphere 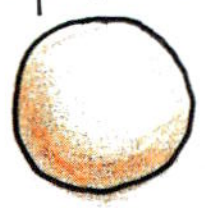	wholeness, protection, strength	roundhouse, travel bundle	
spiral 	growth, stretching space, renewal	reciprocal roof, self help strategy	
radial 	draws us to a centre, dispersal	veg bed layout, mail out strategy	
web 	divides space into zones and sectors	woodland management, computer system, e.g. world wide web	
honeycomb 	packing in space, expansion and contraction	tree planting in waterlogged ground, storage of bottles	
torus 	minimises surfaces, dispersal and energising	tyre design, chi gung breathing	
random scatter 	movement, change, dissipates energy	wildflower seed sowing, flash mob	

Applying a Pattern to your Design

...you can come at it from a number of ways.

For example if you are making paths.

THINK/MULL: What is the FUNCTION of a path?

ANSWER: Transport of people and materials across different parts of a site.

THINK: What energy do I need for these tasks? What would benefit this?

ANSWER: Freedom of movement on main routes, narrower access on less used paths.

PATTERN APP: Check through the patterns sequentially.

THINK: Select appropriate pattern e.g. branching or web.

SKETCH: This pattern onto your design map.

EVALUATE: Walk the pattern on the site as if the feature was there, in this case branching paths. Use sticks to mock up and mark out a realistic layout.

TWEAK: Make adjustments to your plan according to the feedback you received from trying out the pattern.

You may well merge more than one pattern and function, e.g. branching and web, distilling the essence, perhaps without the set form – just like the spider makes a web to fit the space rather than always weaving a perfectly symmetrical pattern.

This is an App.

You download to your body and brain for free. The Pattern App. can then be applied consciously or not to support a function in order to organise energies. It also applies to information and organising people and resources to make your design effective. I have made this App for you, distilled from many books, however your own interpretation of a pattern is perfect and your personal insight from observation and practice will always be uniquely optimum for you.

I am drawn to: pattern,

because it seems to be:

at this time.

Permaculture Principles

Today we are going to take a short walk to see for ourselves if the permaculture principles underpin every wild and thriving ecosystem as asserted by the founders of permaculture. You can do this exercise at your site. However, if your site has very little diversity and natural activity you'll probably enjoy this section more by going on an outing. The coast, the woods or any fairly unmanaged place, such as an unruly mixed hedgerow, works well to observe the permaculture principles in action. In fact, even 'the weeds' growing in a town centre or abandoned urban concrete space will reveal these principles.

For many people, the initial response to the permaculture principles is 'well, that's just common sense.' Bill Mollison's response to that was 'if common sense is so bleedin' common, how come we're in the mess we're in?' And yes, it is true – the principles that underpin all processes of life are by their nature common, obvious and sensible. It's just that we've had a century of 'cheap' energy that led us down a path of being able to override this common sense and impose energy intensive designs in our settlements, industry, trade and agriculture.

Over the decades the permaculture principles have been developed and disseminated across the globe, resulting in a number of versions. We are going to work with David Holmgren's take on it, shown here, as they are particularly accessible. The bold type shows the potential and positive aspect of the way natural, thriving ecosystems work. The text in italics, are all well known folk sayings – that remind us of the limits or cautionary aspect of the principle. This also reminds us that indeed these ideas are, or rather were, common sense. Traditionally it was very important, if not essential for survival, for individuals and communities to respect and understand the LIMITS to consumption of their resources. Hence why they became enshrined in folklore as common place sayings; and a very useful contribution to the permaculture principles for us.

A final point is that "the first 6 principles tend to consider systems from the bottom up perspective of elements, organisms and individuals. The second 6 principles tend to emphasise the top down perspective of the patterns and relationships that tend to emerge from self-organisation and co-evolution." (Holmgren, Permaculture: Principles and Pathways Beyond Sustainability, 2002). This is not vital to understand, though it is interesting when you think about it.

1 Observe and Interact
Beauty is in the eye of the beholder

2 Catch and Store Energy
Make hay while the sun shines

3 Obtain a Yield
You can't work on an empty stomach

4 Apply Self-regulation and Accept Feedback
The sins of the fathers are visited on the children unto the seventh generation

5 Use and Value Renewable Resources and Services
Let nature take it's course

6 Produce No Waste
A stitch in time saves nine
Waste not, want not

7 Design from Patterns to Details
Can't see the wood for the trees

8 Integrate Rather than Segregate
Many hands make light work

9 Use Small and Slow Solutions
The bigger they are, the harder they fall
Slow and steady wins the race

10 Use and Value Diversity
Don't put all your eggs in one basket

11 Use Edges and Value the Marginal
Don't think you are on the right track just because it is a well-beaten path

12 Creatively Use and Respond to Change
Vision is not seeing things as they are but as they will be

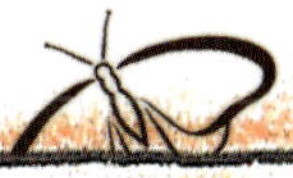

Observing the Principles in Nature

You may take this book or your notebook, pen and pencil and whatever you need to be comfortable to stay outside for a while. Here are some suggestions as to how you can start applying these principles straight away and in any context:

OBSERVE AND INTERACT: commit to observe without making assumptions. Observe the weather and how you are feeling today and plan accordingly e.g. wear a coat.

CATCH AND STORE ENERGY: keep warm, pack some layers; do something else you need to do on the way to your destination.

OBTAIN A YIELD: collect some wild edibles at the same time. Take a friend or a child that would benefit from a trip out. Consider writing an article or blog about the process.

APPLY SELF-REGULATION AND ACCEPT FEEDBACK: go somewhere based on previous experience that you know will be enjoyable, this may mean going somewhere new. If you always end up hating walking too far or on a windy day, apply that info and regulate your activity accordingly.

USE AND VALUE RENEWABLE RESOURCES: walk there, cycle there.

PRODUCE NO WASTE: take a flask so that disposable cup of coffee doesn't tempt you.

DESIGN FROM PATTERNS TO DETAIL: for example, if you have a pattern of being time scarce, combine the trip with other errands or design an outing when time is available.

INTEGRATE RATHER THAN SEGREGATE: do some pattern observation at the same time. Do your shopping on the way home. Take your elderly relative, park them in a cafe whilst you do the exercises if they don't want to join in.

USE SMALL AND SLOW SOLUTIONS: go somewhere simple- don't plan a trip that will sap your energy.

USE AND VALUE DIVERSITY: keep the principles in mind in different settings. Use the journey, your destination today and your own back garden or street at other times. Include friends, colleagues, students in the exercise.

USE EDGES AND VALUE THE MARGINAL: go somewhere relatively untamed or unusual for you, or do the exercise on the way to somewhere else. Perhaps, put it to the test by going somewhere you have to go even though you don't feel like it.

CREATIVELY USE AND RESPOND TO CHANGE: if this is new territory for you, that equals change, work with it and your reactions consciously. If something comes up you didn't expect, see if you can creatively adapt to the situation.

DOES THIS GIVE YOU AN IDEA OF HOW WE WILL USE THE PRINCIPLES...?

SO, ARE THESE PRINCIPLES REALLY UNIVERSAL? LET'S CHECK...

Remember: we are looking to observe these principles in a natural setting in order to name something actually happening or visible. Sometimes people name ideas or solutions that relate to the principle, like my suggestions opposite; that is the NEXT step on, applying them to our actions, in other words designing with them.

So bear that in mind, as our aim is first to AUTHENTICATE these principles as observable and sensible, in order to TRUST them as guiding characteristics in our design work.

NO WORRIES ALERT

Your personal interpretation or association with the wording of the permaculture principles is plenty good enough. That's the beauty of them, whether you are a Professor of Earth Science or a complete beginner, they have relevance.

There is a plethora of excellent books out there unpacking the permaculture principles in detail. My task here is to accompany you on your personal investigation

At your chosen destination, drop into your observation mode by softening your vision, coming present to your senses and the feeling of relaxed alertness. From the following pages, select a principle and let your attention wander over the place/space to discover in what form or process you can see it at work. If you notice tension in your body or disturbances in your mind, such as irritation or wandering thoughts, that feel like they are hijacking the exercise or rendering it impossible, then there are a couple of things you can do.

You can magnify the thought/sensation and see where it leads you. For example, you are thinking, 'I am cold, this exercise is boring': Repeat the internal voice deliberately, make it louder, really feel where you are cold, how bored you are and the sensation of that.

Or, you can express it in another sensory mode or channel. Translate the cold or the exasperation into a movement, a sound or an image either scribbled in your notebook or in your mind's eye.

Then, bring one of the permaculture principles to mind, let your attention slide sideways and see what catches your awareness in the place/nature around you., e.g. a leaf, the horizon, litter, anything. Then unite this with the disturbance or irritation you were feeling. It is likely you will see some correlation between the disturbance, the permaculture principle and the place around you and with that gain a personal insight that will resolve the irritation and let you carry on the exercise.

Note in the bubbles overleaf how, what and where you observe the principle in evidence in the wild around you or within you. Scribble or sketch a visual prompt on the page as well.

Note your observations in the bubbles, if you get stuck, help is at hand on the following pages...

OBSERVE & INTERACT

"Beauty is in the eye of the beholder"

CATCH & STORE ENERGY

"make hay whilst the sun shines"

OBTAIN A YIELD

"you can't go to work on an empty stomach"

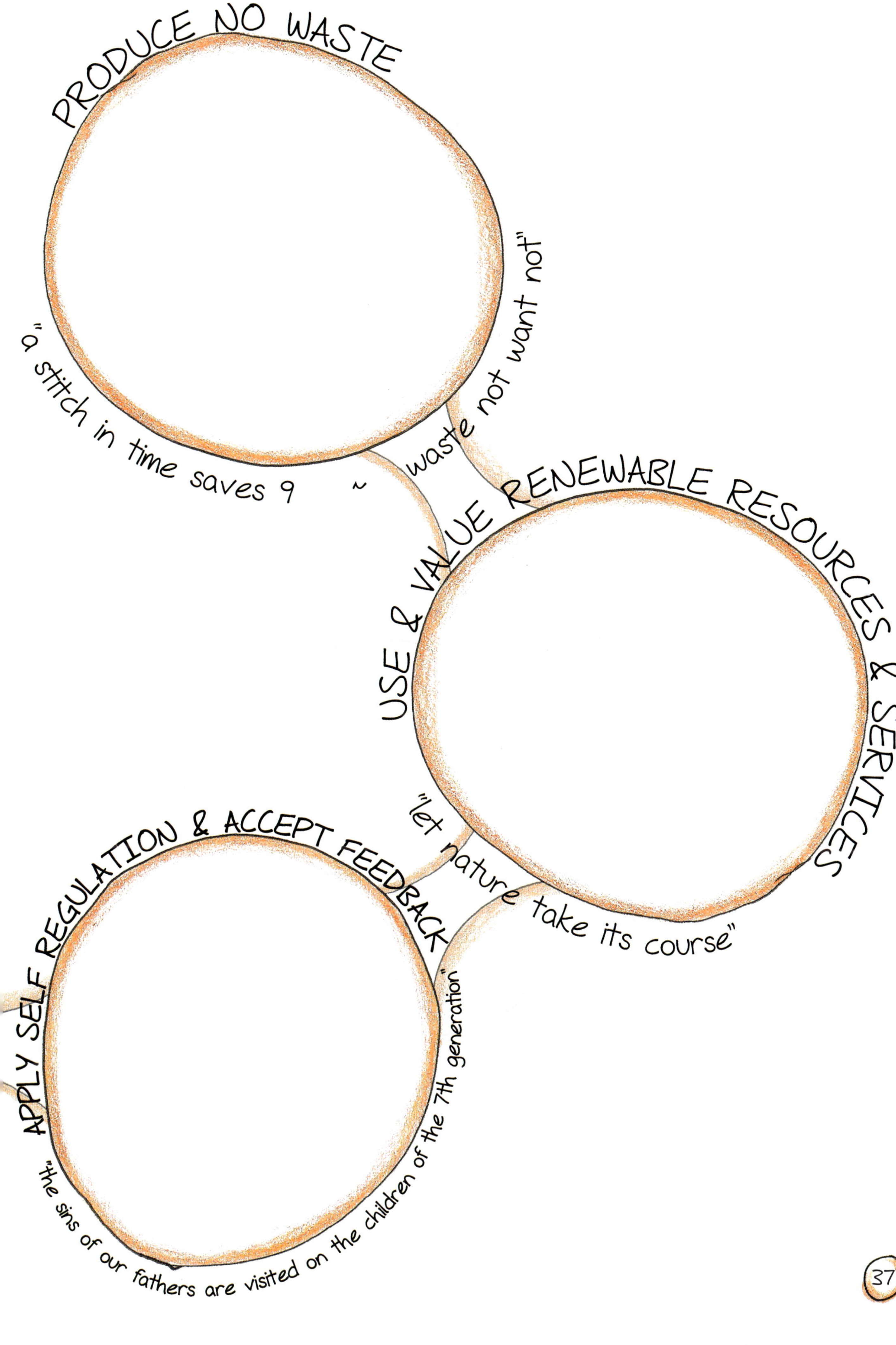
PRODUCE NO WASTE
"a stitch in time saves 9" ~ "waste not want not"
USE & VALUE RENEWABLE RESOURCES & SERVICES
"let nature take its course"
APPLY SELF REGULATION & ACCEPT FEEDBACK
"the sins of our fathers are visited on the children of the 7th generation"

DESIGN FROM PATTERNS TO DETAILS
"can't see the wood for the trees"
INTEGRATE RATHER THAN SEGREGATE
"many hands make light work"
USE SMALL & SLOW SOLUTIONS
"the bigger they are the harder they fall ~ slow & steady wins the

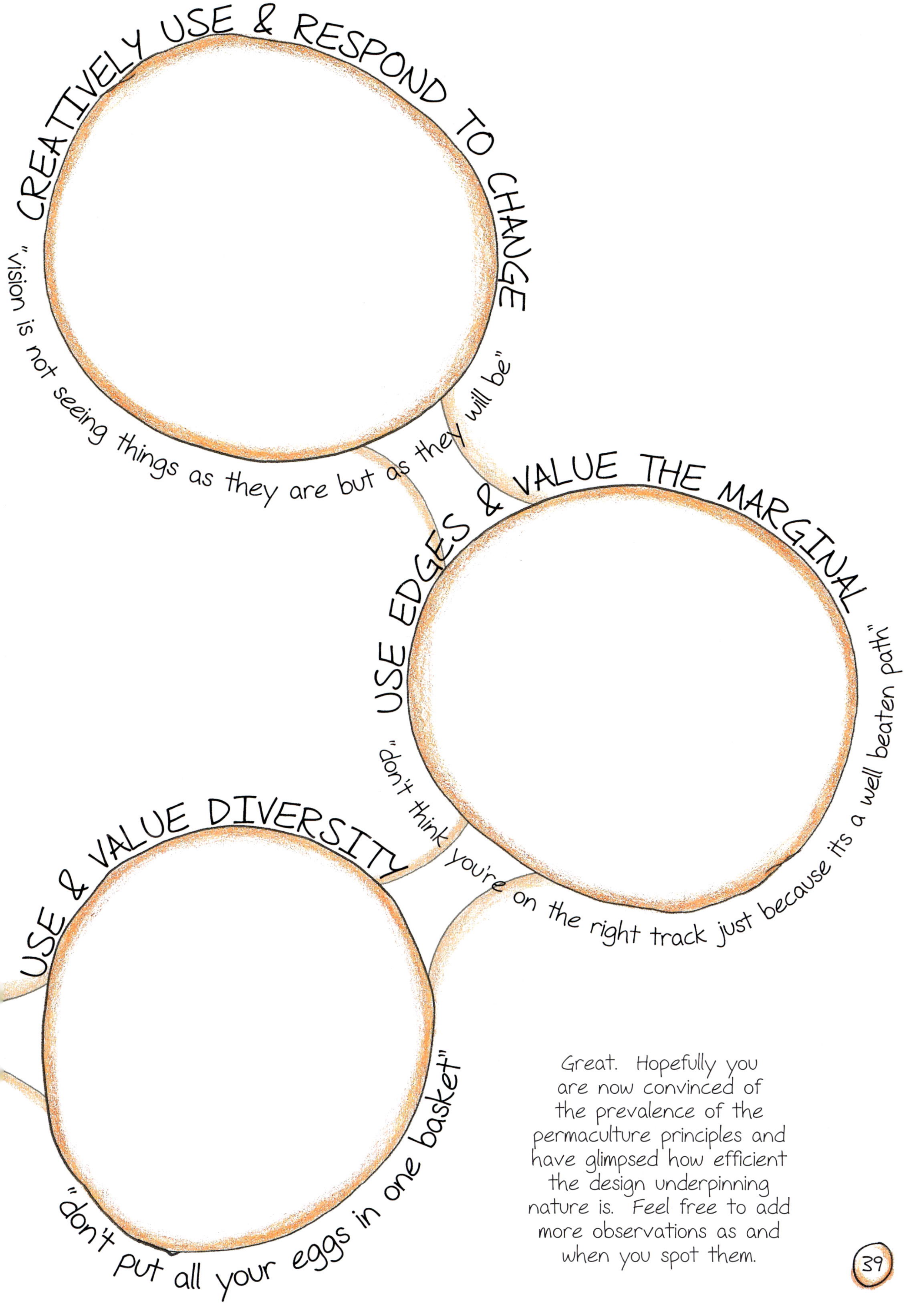

Great. Hopefully you are now convinced of the prevalence of the permaculture principles and have glimpsed how efficient the design underpinning nature is. Feel free to add more observations as and when you spot them.

Applying the Principles in Design

Now although we haven't got into our formal Design Process yet, it's worth spending half an hour or so applying the permaculture principles to our own design while they are fresh. This is a casual exercise to stimulate and capture ideas. We will revisit them again in more detail as a Design Tool after we've completed our surveys.

PRINCIPLE	IN NATURE	IN YOUR DESIGN
EXAMPLE integrate rather than segregate	gorse, nettles, thistles growing together to protect soil, fix nitrogen, provide habitat for birds and food for insects, retain water and more.	e.g.: combining big jobs like site preparation or tree planting with social activities; volunteer help or local educational group OR planting shrubs on edge of trees to diversify habitat and slow down wind EXAMPLE
observe and interact 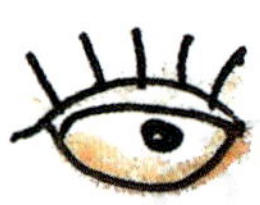	smell activates saliva which produces enzymes for digestion of food. Plants will 'observe' an available space and then 'interact' by growing into it.	
catch and store energy	branching pattern in trees, their leaves and their roots optimise capture and storage of nutrients, energy from the sun and water.	
obtain a yield 	leaves capture sunlight and turn it into food; fallen leaves and bird poo become compost and feed surrounding life.	
apply self regulation and accept feedback	a plant or tree grows according to what soil, light and shelter is available.	
use and value renewable resources and services	sun, soil, air, water powers all growth, incorporating any 'waste' into useful organic matter.	

PRINCIPLE	IN NATURE	IN YOUR DESIGN
produce no waste	all organic matter is recycled by biodegrading; a slug or a fungi can even be seen breaking down old plastic. No 'output' is wasted or left to accumulate and become pollution.	
design from patterns to details	branches, shells, leaves and clouds. Zones of growth depending on light and shade in a hedge, coastline or hillside.	
integrate rather than segregate 	gorse, nettles, thistles growing together to protect soil, fix nitrogen, provide habitat for birds and food for insects, retain water and more.	
use small and slow solutions	growth over time; a tree or river valley evolves slowly; a tiny plant grows in a crack in the pavement to test conditions, a pioneer first rather than a fully fledged forest.	
Use and value diversity	numerous plants occupy a small space, maximising availability of light and nutrient over the course of a season and over years.	
use edges and value the marginal 	'weeds' colonising new ground, accumulating minerals for all; hedgerows full of life on the edge of a bare field; seagulls nest on sheer cliffs to find shelter, proximity to food and undisturbed nesting sites.	
Creatively Use and Respond to Change	a bare patch will quickly fill with plants; soon joined by a pioneer tree like willow, succeeded by oak or similar, with a life span of centuries. Coastal life adapts to changing tides; coastlines are dynamic responses to longer term processes.	

PART 2:
Seeing Things Anew

Design Processes

Managing Complexity

We learnt earlier in the book that permaculture differs from organic gardening by addressing every aspect of human needs, particularly placing the importance of a DESIGN PROCESS at the centre of its meaning. What is a design process? – well, put simply it is a series of steps for being effective and managing complex tasks. In particular, it prioritises and gives hierarchy to these steps – in other words, using a sequence that builds up in stages. For our purposes, it means a thorough observation and survey of a site and the people involved before making decisions. We're going to use a design process known as SADIMET and have a glance at a few other ones you may be interested in.

My mate was in the SAS and then a bodyguard to famous rock stars. He now stewards his 7 acre smallholding with permaculture and noted that the design process he used in his old career was almost identical to the frameworks adopted by permaculture. The emphasis as a soldier and bodyguard was to constantly observe and interact with the environment at each stage of an event or project. This is a helpful reminder for us to have constant vigilance and keeping our peripheral vision open whilst designing – rather than holding rigidly to an idea we are now solely 'analysing', 'designing' or 'evaluating.'

When we become fluent at a task, our design process becomes second nature and we use it intuitively and seamlessly like a skilled surgeon or an experienced cook. Then we are in a state of 'relaxed alertness', effective and competent. Cooking is a classic example most people will relate to –

we OBSERVE AND SURVEY – how hungry are we; the ingredients in the cupboard; the time available and the preferences of the other eaters.

We ANALYSE this information, consider pros and cons of our options and then make a...

DESIGN decision e.g. to cook macaroni cheese.

Seamlessly we IMPLEMENT this design by selecting the appropriate pans, utensils etc.

We MAINTAIN the process by observing and interacting constantly with our senses – the aroma, taste, textures, heat of the hob, sound of bubblings and so on. This happens whilst also taking account of external factors, such as children in the next room; television or radio, time and other matters on our minds.

We are evaluating and applying feedback of that data as we cook and as we do afterwards when we either think, yum! Or yuk! A recipe that worked or a flavour that did not, is instantly EVALUATED AND TWEAKED into our future cooking.

I use a design process when I am:

The steps I follow are:

<u>tick the boxes that apply to you:</u>

I do this intuitively	☐	without thinking
With careful reference	☐	to the steps
And skip steps	☐	with mixed results
I don't recognise such a process	☐	and it's all fine
I don't recognise such a process	☐	and it's all a mess

Whether formal or informal, somewhere you are using a design process

E.g. looking after children, pets, elderly, plants, or my own health

E.g. playing an instrument, singing, gardening, drawing, sewing, carving

E.g. DIY, mechanics, engineering, IT, interior design, studying or teaching

SADIMET Design Process

SADIMET is a design process framework, commonly used in permaculture and more mainstream industries. From this point on our work will be following these steps.

SURVEY: 'what are the basic parameters to work with'

Design brief and vision: a statement of intention and desired outcomes.

Landscape or site survey: biodiversity, resources, slopes, sectors (wind, sun, water, soil), access, structures, views etc.

Base map: scale drawing of all existing features.

People/ client survey: needs, desired outcomes and yields, time/energy, skills etc.

Observation: 'maximum contemplation minimum action'. We've made a start on this practice and a commitment to use it throughout the design.

ANALYSE: 'computing it all from every angle'

Key Planning Tools: zone, sector, elevation and network analyses are used to harmonise our vision and resources with the landscape it sits in.

Apply other 'thinking tools' such as Input-Output Analysis to ensure all the elements of the design work together as a beneficial system. Build in resilience and efficiency by working with the relationship of every element and every function within the design.

Research: books, internet, visiting places, catalogues etc...

DESIGN: 'bringing it all together'

Now our ideas have germinated into a rough design, any tensions or contradictions can be resolved by returning to the thinking tools.

Use to scale paper templates to represent proposed trees, structures, plants, ponds etc. Keep moving them around on the base map until it all clicks into place, checking distances and relationships between the elements, shading, windbreaks etc.

Apply permaculture principles to check every angle is covered.

Draw a conceptual or artistic map of your decisions. Draw a 'to scale' design map to match the original base map.

IMPLEMENT: 'make it so'

Make a realistic time schedule of which order and when you plan to put the design in place.

Create materials lists, costings and methods for achieving your design.

Decide who is doing what and if extra help and skills are needed.

The above are flexible, however a clear plan will help tasks be manageable and therefore better achieved.

MAINTAIN: 'keeping gain without pain'

A practical plan of how to manage the above physically, mentally, financially and so on.

Trouble shoot where assistance or other people's input may be desirable or imperative.

Think how you will sustain tasks whilst new ones need attention.

EVALUATE: 'honest check in'

Over time, assess the effectiveness of the project as it unfolds.

Keep records of what worked and what didn't and unexpected results.

TWEAK: 'keeping it fun, real and effective'

Adjust your activities or design layout to include lessons learnt to create useful positive feedback loops.

Regularly evaluate what you're observing and learning, improving resilience and reflecting the dynamic processes and cycles at work!

Exploring Complexity

overwhelm and the value of a design framework

Even in a small, simple site, if we respect nature, our own depths and the feelings of those around us, we soon see just how many dynamics are to be accounted for. Luckily our brains are super computers and can cope. However, overwhelm or program crash, is also possible. This can lead to problems and system errors. A design framework breaks down vast quantities of info and allows it to be digested, just like a piece of software or operating system in a computer.

Here is a 'thought-fest' exercise (formerly known as brainstorming) to illustrate the multitude of factors at play in a given situation that could impact on your design. There is a page at the back of the book (Managing Complexity) with suggestions of what you could consider in the realms of land, society and the personal. Cast your mind wide and note key words that may apply to you in the relevant sections on the next page; ask a friend, colleague or family member if they can think of anything you've missed.

Is ignorance bliss?

Can we ignore the type of factors we just articulated? Perhaps. However, as permaculture designers we are aiming to be conscious of as much of the whole system as possible; given the understanding that all factors are in relationship, influencing and impacting our design and that to block or circumvent them will require energy in some form.

This exercise reveals the tip of the iceberg of the complexity at work in a landscape, society and what we personally bring to the table. Combine those 3 and we have major complexity and that can lead to overwhelm, skipping pertinent issues to reduce stress or a complete unawareness the factors exist at all. The beauty of a design process is that it frees us from this pressure and these concerns. A step by step approach, in the right order, gives our brains space to assimilate such dense information. The process reduces this complexity to manageable chunks and our supercomputer brains do unseen work effortlessly, leading us to sound choices.

Even if we don't create 'perfect' designs, the use of a design process and a coherent thinking/action framework can reassure us that we have done our best given the resources available. By starting small and close to home, we also gain experience, have space to incorporate feedback and minimise wasting our time and energy.

We are going to use SADIMET for the rest of the book. However feel free to invent your own frameworks or adopt another. On the following pages are 3 other design frameworks commonly used in permaculture, industry, design and management. There is a comparative summary of the design frameworks in the back of the book.

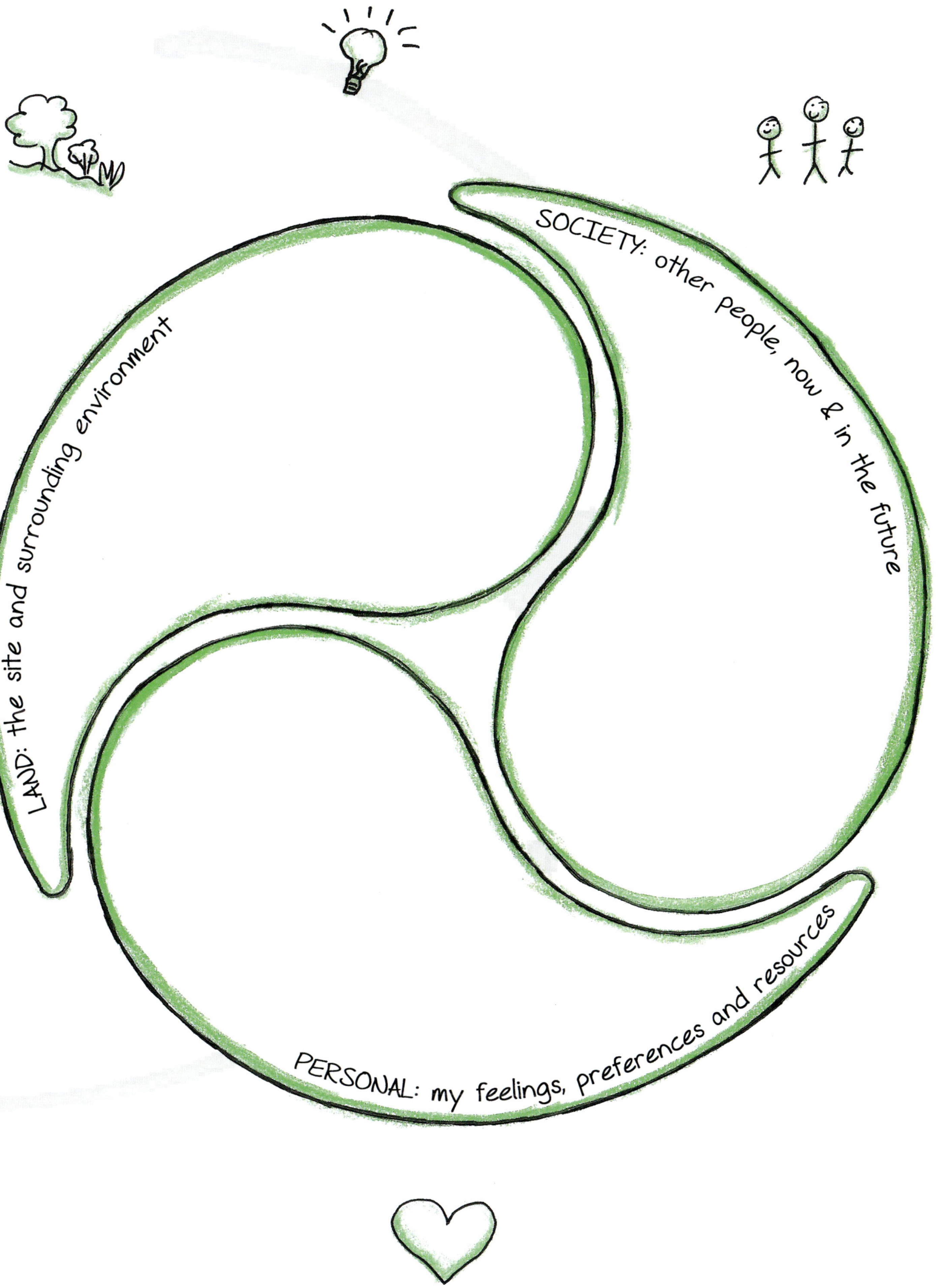
LAND: the site and surrounding environment
SOCIETY: other people, now & in the future
PERSONAL: my feelings, preferences and resources

Other Design Processes

CEAP

COLLECT:
info about the place, the people, yourself, funders, related projects, relevant paradigms etc.

EVALUATE:
relevance, priorities, suitability of ideas, their placement, sustainability over time etc.

APPLY PERMACULTURE PRINCIPLES:
see if you have met all the characteristics of resilient and efficient systems in your design.

PLAN:
timescales, budgets, out-sourcing jobs and roles; contingencies for possible challenges.

Appreciative Inquiry

DREAM:
allow yourself to envision your ideal scenario and outcome for the project. Use creative techniques to expand this and then name clearly your intention for the design.

DISCOVER:
investigate yourself, the place, current processes and methods being used. Explore related topics; visit other projects; research your ideas and alternatives.

DESIGN:
consolidate the above into key elements with beneficial relationships between them.

DELIVER:
make a plan of action with attention to timings, budget, mode and method of implementation.

\+ Positive:

OBREDIMET

OBSERVATION:
of place, people involved and yourself.

BOUNDARIES:
physical and psychological, how they interact with your project and their implications.

RESOURCES:
survey skills, resources and materials available on the site and available to you from other networks and sites.

EVALUATE:
discern relevant design priorities and how they fit your vision, site and resources.

DESIGN:
assemble key elements spatially and be clear why they are included.

IMPLEMENT:
work out what activity happens when and how and by whom.

MAINTAIN:
check for realistic and achievable methods of keeping the design in progress.

EVALUATE:
see what's working, what isn't and what to do differently to improve the process.

TWEAK:
adjust what you are doing and how you are working to feedback what you have learnt from the previous cycle of activity.

- Negative:

What did you find Positive, Negative and Interesting in this chapter? Consider the purpose of using a design process and the different frameworks.

? Interesting:

Vision

Having a clear and inspiring vision is a condition of later success in your design. The vision does not need to proscribe in stone what you wish to achieve; more accurately it is the spark of life that keeps you on track. A clearly articulated vision enables you to access relevant, effective and enjoyable design solutions as we proceed. It is totally flexible and appropriate to be tweaked as you go along, whilst retaining its role as a reliable guide. A pattern to inform detail as you progress.

Colour and imagery stimulate the senses and thus catalyse the brain, activating enthusiasm and articulating more than words often are able to. To that end our next task is to create a Vision Board, like the example to the right. This process, of selecting imagery concurrent with your wishes and aspirations for the design, can often bring a sense of ownership and confidence to a project, especially if you feel overwhelmed or unclear about what to do next.

Making a vision board also allows us to daydream and loosens the bonds and critical voice of our unconscious minds that can inhibit clear analyses or creative thinking. Every great thinker and inventor has used some sort of visioning process to birth their big ideas from Leonardo di Vinci to Steve Jobs. Success books and gurus frequently cite the likes of Richard Branson, Einstein and Walt Disney for the weight they placed on this early visioning stage and the importance of letting the mind wander, dream and think beyond the realistic. Stretching this edge fits with the permaculture principle of valuing the marginal and walking the road less travelled.

Frequently we are constrained by what seems possible, reasonable, inoffensive, appropriate to our status and so on. If we dare to dream and can go beyond established norms we will surprise ourselves by the creative forces unleashed and can count on that old cautionary inner voice to keep us to our limits.

"Our deepest fear is not that we are inadequate.
Our deepest fear is that we are powerful beyond measure.
It is our light, not our darkness that most frightens us."

Marianne Williamson.

Designers, inventors, entrepreneurs, dictators and the advertising industry have long understood the Power of the Imagination activated by colour, imagery and vision. Educators also are coming to realise the importance of harnessing creativity to further our analytical thinking, often known as Brain Based Learning. So, as humble permaculture designers, I encourage you to give as much respect to your vision board and other arty exercises, as you do to your number crunching and research. Permaculture has a long tradition of combining the creative and the analytical, as befits holistic, systems based thinking. We can combine the different parts of our brain to come up with emergent solutions; just like nature combines hydrogen and oxygen and get something as miraculous as water.

EXAMPLE OF A VISION BOARD:

Midway through the implementation of my smallholding design, I made this vision board to help anchor my desired balance of work and pleasure.

Benefits of a Vision Board

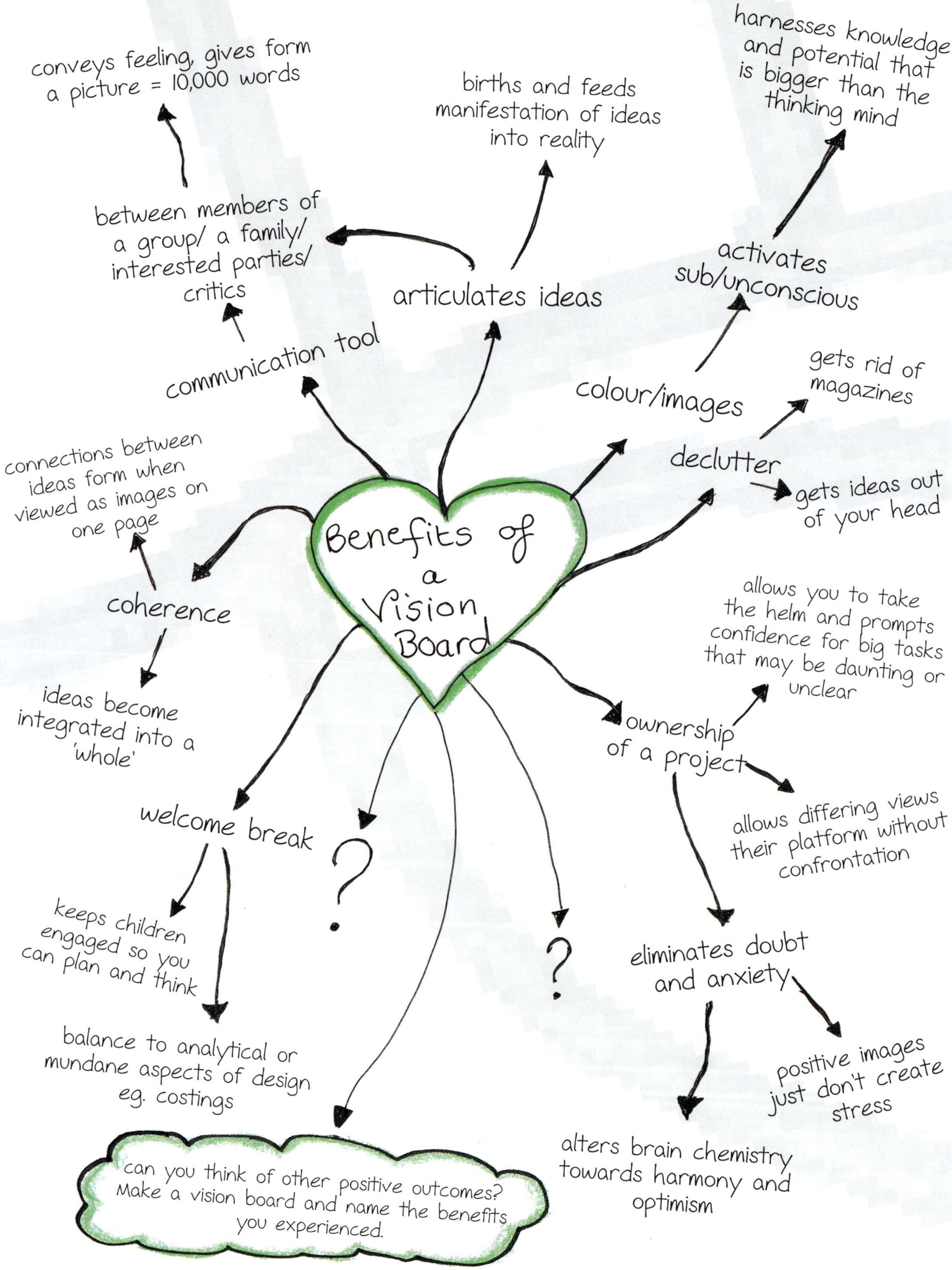

I have done this exercise with tons of people and vision boarding always stimulates enthusiasm. The only exception was an MI5 officer; although he did at least cut out a picture of a shire horse pulling a cart of home made cider. Vision boards breathe meaning into a project and generate ideas, that's why blockbuster films such as Harry Potter devote huge resources to this practice.

If you have reservations about making collages, art or similar not being your thing, I'd like to reassure you that nowadays, even hard nosed executives and tough blokes now have fun and find value in a vision board.

The following disclaimer is a release form:

Airy-Fairy Disclaimer

Participating in the creative process of making a vision board is an experimental process. It does not affect your statutory right to be a serious, professional person, or affect your status or image in any way.

There is no right or wrong result and if appropriate you can consider the exercise as humouring the author.

Participant signature: My signature: Dale

This form protects your reputation, seriousness and any other cherished characteristics that may be compromised by making a vision board.

Other ways to develop your vision include:

- [ ] film (capturing images yourself or watching films for inspiration)
- [ ] visiting places that evoke the feeling or outcomes you wish for
- [ ] right brain activities leading to insights e.g. walking, cleaning, cooking, weeding
- [ ] daydreaming, night dreaming, imagining how you would like the finished design to look and feel; sounds and scents you would like to encounter

tick

Making a Vision Board

All you need for this exercise is glue and a few magazines, even scissors are optional as some rough tearing out of images is in itself a freeing up experience from the rigours of perfection and form!

There are 2 main approaches to making a vision board, I like to combine both:

~ finding specific images that you wish to include in your design

~ collecting up any images you are drawn to whether specific to your design ideas or not.

Some people like to do this activity in a leisurely manner over time; on my courses we usually give about 2 hours to make our vision boards and people add more to them when they return home. 'The Artist's Way' suggests another good method of getting a minimum of 10 magazines and allowing a mere 20 minutes to flick through them tearing out any image that catches your interest strongly.

It is best practice to have your vision board visible throughout your design process, somewhere it will catch your eye regularly. If you are working with a group or on a family project you can make one together or produce individual ones. This will be fertile ground for discussion, revealing similarities and differences in your vision and greatly informing your design. For example, when I was designing our bathroom, our vision boards revealed the following: my son chose all black modernism; my daughter vintage chic; my husband a jungle theme and my own preference was for a Tuscan hotel aesthetic.

It is better to know differences in vision early on. Convergences can then have space to arise, creativity will emerge. If suppressed or glossed over, differences and misunderstandings will inevitably show up down the line and may well be harder to reverse or amend. My greatest regret as part of an eco-village for 10 years was not flagging up the centrality of a shared vision at the start. At an early stage of forming the group, someone wrote a well meaning vision statement independently. I knew from my training that a group, explicit process was fundamental to good design, however I was shy and inexperienced and put my visioning session notes back in my bag and said nothing. A decade later, after many misunderstandings, I have learnt my lesson well: explore difference, embrace diversity or be prepared to accept the feedback of the unexpressed when it finally finds its voice.

By its nature, a Vision, as the first step in the design process, is liable to change as you find out more and more about the people and place involved and so is flexible. The imagery you choose for your vision board is not sacrosanct and can be changed, even scrubbed out. Potential resistance will also show up, the Inner Voice of Niggle or perhaps the views of a relevant other person. This is better expressed at the outset and explored than repressed or ignored. So to reiterate, joint vision boards, especially in a community group or concerning a public activity, are a great foundation to work from.

MAKE AN A4 VISION BOARD HERE

make a bigger one as well for your home/work place/group meeting venue

use a flattened cardboard box, a 3D box or even on any object of your choice

use lots of colour and fill in ALL the gaps

Vision Statement

Your vision statement is the partner of your vision board and also is best to display somewhere you will see it regularly, where it can be tweaked and digested over time. Remember all those parked ideas in the Observation chapter? It's now time to let them loose.

Here are some pointers for your statement:

~ 150 words maximum

~ name the process of how to achieve your design, e.g. fast; without stress etc.

~ reflect a good sense of the desired end result

~ include specifics or name a general approach or both

~ keep it short, sweet and simple

Bullet point up to 10 key elements that you expect to be in your design:

eg. a pond, a hammock or a celebration

Evaluation

+ POSITIVE ~ I enjoyed:

because:

- NEGATIVE ~ I found it challenging to:

because:

? INTERESTING ~ Unexpected or unusual outcomes were::

Landscape Survey

Whether you are designing for a small garden or a large piece of land, a landscape survey is essential. Remember we are still refraining from imposing set ideas, even though our vision for the design is now defined. The key to a design that is truly in harmony with the landscape is doing as thorough a survey as possible. Huge potential lies in any landscape to harness energy from existing features to meet the needs of our vision, minimising our use of fossil fuels. For example, a slope creates gravity to move resources such as water without pumping, or to move materials without the car or machinery. Orientating elements, whether structures or gardens, in relationship to the slope and sun direction, captures huge amounts of heat and shelter which otherwise might create more work, lower yields or need more fuel to sustain their performance in some way.

We are going to use a trusted landscape survey; some of it may not be relevant, though I urge you to fill in as much as possible – you never know when a piece of landscape information may become relevant later on. By completing the survey you are doing your best to ensure the complexity of the land is being accounted for.

You may add info to this survey over time, as with your observations of patterns and permaculture principles. Remember, there is no need to crunch all this data and its implications instantly. Just by collating it, your mind will do a lot of processing effortlessly. In Part 3, we will go deeper into landscape observation combining it with analysis as we start to find the most appropriate locations for all our design elements.

Excellent sources of data for the survey include:

~ local people/ prior owners/ neighbours

~ MET office

~ local geology and botany societies

~ research through books/ library/ internet

Companies, such as Groundsure, provide good surveys on scale maps of potential flooding, pollution, contamination, underlying bedrock and so on. If your design is for a large development, a house or an ecologically sensitive area, it is well worth the £50 for this data.

A copy of the landscape survey is in the Additional Resources at the back of the book for making copies for other people and future projects.

SURVEY OF:	DATE:

EXISTING BIODIVERSITY

ecosystems (eg. hedgerow, marsh etc.) plant species animal species (include pests) old trees what's growing... ...how well	

LANDFORM

elevation (metres above sea level) slopes (erodibility / condition) aspect (which way slope faces)	

MICROCLIMATES

frost pockets sun traps shady places windy places sheltered places damp / wet places well drained / dry places combined effects	

SURVEY OF:	DATE:
CLIMATE	
rainfall (mm/year) average temperatures first & last frosts prevailing wind patterns in different places	
SOIL	
pH/ nutrient tests indicator plants profiles texture variations	
WATER	
<u>SOURCES:</u> mains existing taps roofs and downpipes greywater springs streams ponds possible pond / water storage sites  <u>NEEDS:</u> irrigation (crops/garden) animals aquaculture domestic (drinking, cooking, washing) commercial (e.g. food processing)	

SURVEY OF:	DATE:

SERVICES

underground overhead cables sewage systems land drains mains electric off-grid electric	

STRUCTURES

<u>BUILDINGS</u> condition energy potential (solar, rainwater etc) relationship to land (past and current purpose) <u>WALLS & FENCES</u> aspect condition	

ACCESS

<u>PATHS, ROADS, TRACKS, GATES, BRIDGES, DOORS ETC</u> siting type & frequency of traffic steepness/ erosion potential condition seasonality desire lines (where folk naturally walk)	

SURVEY OF:	DATE:
BOUNDARIES & INTERACTIONS	
north	
east	
south	
west	
BEAUTY	
views from site	
within site	
off site from outside	
MISCELLANEOUS	
resources	
hazards	
special areas (e.g. archaeological / sacred)	
off site services (eg transport/ markets)	
unique features	

Other Landscape Survey Resources

QUADRANT SURVEY

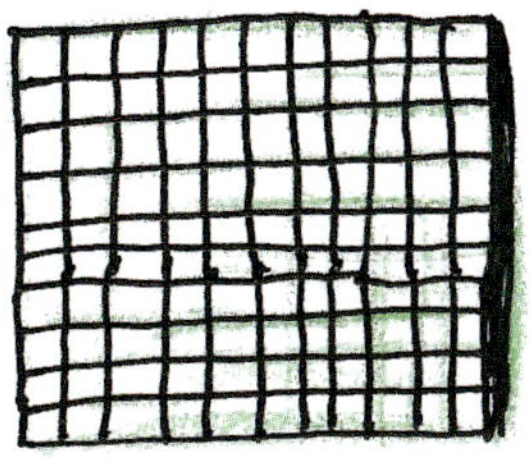

If you are interested in detailed data a 'quadrant survey' is a good idea. This entails getting or making a grid of 1m x 1m and then dividing it into 10cm x 10cm squares. You then place the quadrant grid on different areas of ground and record numbers and species of plants in any given square in the grid. This is only useful from spring to autumn as most plants will be underground in winter. A quadrant survey is valuable if you wish to demonstrate a very thorough and professional survey for permission to carry out a design, especially in an ecologically sensitive area. Always err on the side of caution with regard to what may already be growing on your site.

FIXED POINT PHOTGRAPHY

A further resource in your landscape survey is to begin a practice of Fixed Point Photography – as the name implies – select a few fixed points from which to take a photo of your site, perhaps at every corner. We selected 4 points covering the main areas of our land, taking a photo from these points every season, lessening to once a year as shrubs and trees grew obscuring visibility from each point. Fixed point photography is an excellent record of the changes you will be making and for understanding the pace at which your design grows. It is also good in relation to funders, planning authorities and neighbours, both as proof of improvement of the site and/or as evidence of activity going as planned. After a few years it can be very hard to perceive changes that you've implemented and the evolution of a site. Fixed point photography records are also fascinating for yourself and other interested parties to look back on.

March 2009

May 2012

October 2016

Listening to the Landscape

Next we're going to do an observation exercise devised by Chris Day, called Listening to the Landscape. Chris is an important natural architect out here in West Wales, his contribution of sensitivity to and awareness of landscape has birthed many beautiful buildings and inspired people all over the world. I highly recommend his books if you are planning a building in your design.

For this exercise, you'll be using your trackers 180° peripheral vision and dropping into your sensory awareness state. You will also require a notebook or this book to record your observations in the boxes below.

The exercise covers 4 modes. I recommend giving at least 20 minutes per mode on at least one occasion. You can use it over the whole site or to hone in on specific areas, or both! Similarly, you may enjoy to repeat the exercise over time, particularly to tune into changes you are making once we start to implement our designs. After 9 years at my smallholding I still find it useful to do this exercise when tweaking my design at the end of a season.

INTUITIVE MODE:

first impressions ~ walking the boundaries and noting interactions on these edges ~ tuning into gut feelings

OBJECTIVE MODE:

observable data/ facts e.g. it is raining; nettles grow here; the wind drops by the shed ~ you can add this info to your landscape survey and later to your base map

IMAGINATIVE MODE:

the natural tendency of this place ~ imagine this place prior to human habitation ~ imagine this place far into the future

SUBJECTIVE MODE:

go to a spot you are drawn to, either positively, negatively or neutrally do something spontaneous or creative ~ or just do the first thing that comes to mind

Mapping

Alongside your landscape survey, a BASE MAP is vital. It helps you plan, see the bigger picture and communicate about your site and your design with others. By just having it around, the map cultivates familiarity with your place.

A Base Map shows EXISTING FEATURES only. Any design proposals and ideas, thoughts and so on must remain on spare copies of the map.

We will cover some simple techniques for making an accurate map, however Aranya's book, 'Permaculture Design' and Patrick Whitefield's 'Earth Care Manual' are excellent resources for more in depth support in mapping.

Firstly, if you own the site you may well have an accurate, to scale map with your ownership deeds. This will save you lots of time. Alternatively you can look at an Ordinance Survey (OS) map online or in the library and print out an A4 copy. It's useful to include a bit of the surrounds too.

If neither of these is available or relevant, e.g. if you are designing your Gran's patio, then you will need to make your own from scratch.

The beauty of a map is its ability to convey lots of information in symbols and pictures – simplifying quantities of data to a readily assimilated form.

THE ESSENTIAL POINTS TO INCLUDE ON YOUR BASE MAP ARE:

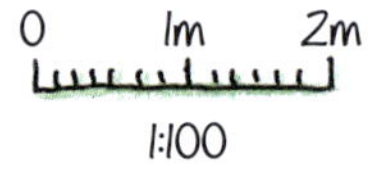

- a scale: e.g. 1cm = 1m, show this with a scale bar in the corner
- aspect: i.e. which way the site faces relative to the sun, show this with a N arrow orientated to the top of the page

- title: basemap; name of place; date
- existing features: make a key to explain the different symbols and demarcations on your map

e.g.

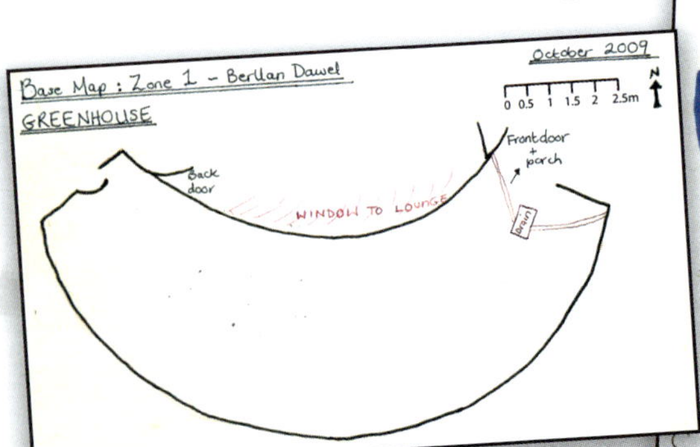

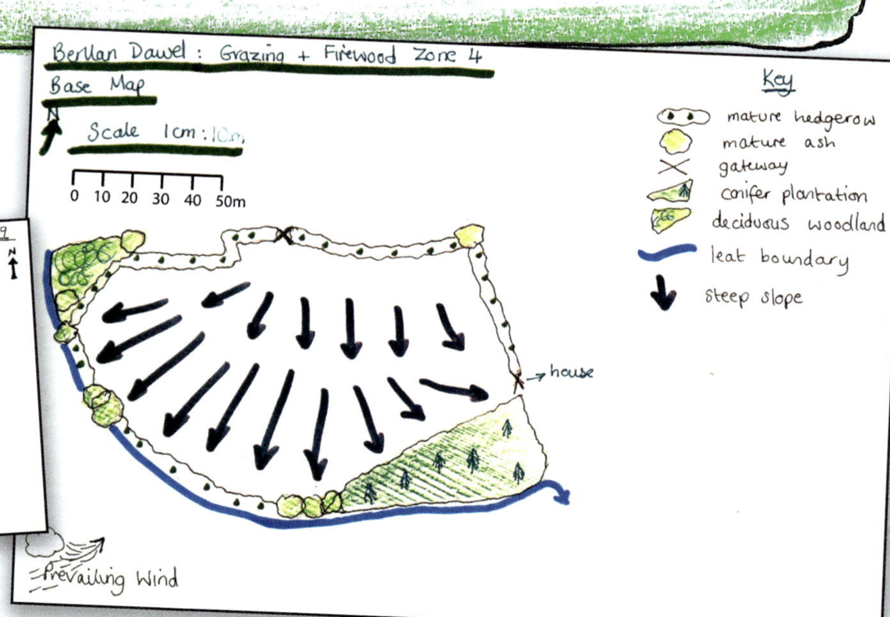

Stick your A4 base map copy here with north pointing up

Use (some) colour

Keep it simple and clear

Mapping Techniques

Next make a few photocopies of your map for experimenting with ideas and concepts; these are also useful for taking outside with you to add data that you can record neatly when you return home. As you come to work on particular areas, whether you are working at a garden scale or on a much bigger site, it is very handy to enlarge sections of the map for noting measurements, possible locations of design elements and other observations.

In addition to an outline of your base map OR if you do not have an accurate map of the site as it is now and need to make one from scratch, it's necessary to have a method to accurately pinpoint where elements (e.g. trees, buildings, entrance...) are WITHIN the map.

2 simple methods are:

CHAIN SURVEY

This requires a long rope or chain. First lay it across the site, anywhere in a straight line works though it is logical to do it on the longest axis along the rough centre of the area you want measurements for. Then with your arms out straight, perpendicular to the chain or rope, establish a point on the boundary that is 90° from the chain. Keeping the line of sight that you've established at a right angle, walk from the chain to the element, measuring as you go. This way you can establish the exact position of everything from this one set parameter

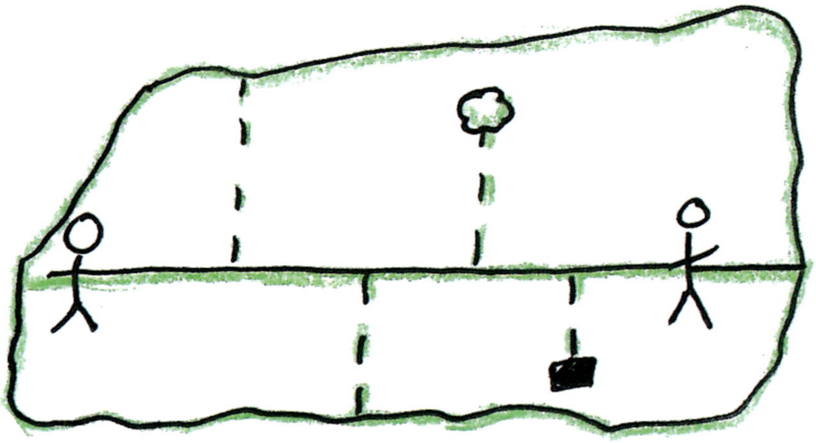

TRIANGULATION

Find 2 fixed points that are easily identifiable (A and B), find the distance between them. Then measure to the 3rd point (C) you wish to find the location of from A and then from B ~ like so:

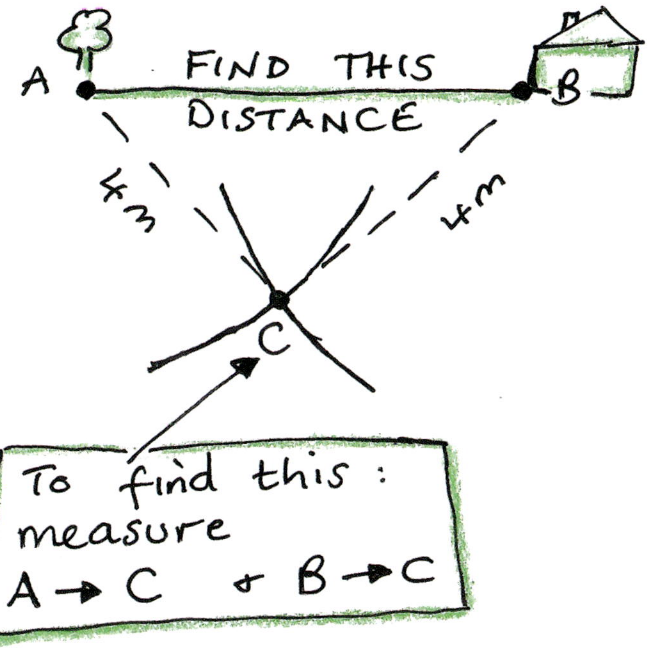

Then on your base map set a compass to the length of those two distances (e.g. you measured 4m from A- C and from B – C) and your scale is 1cm = 1m, so you set the compass to 4cm. Make a pencil arc from point A. Repeat from point B. Where the two pencil arcs meet is the accurate location of the element and can now be marked with confidence on the map.

Attempting to do this by eye is very difficult, as the eye always adjusts angle and distance to suit what's known as 'good form' and squares up the apparent location to suits its own eyedea of what looks right.

Pacing

Using your own pace or stride is a 'good enough' method of measuring, accurate within 5-10% of a a tape measure. Pacing is very handy as you have always got your body with you and can be used in places where a tape measure can't go e.g. over rough terrain, when nobody else is there to hold the other end, or for long distances. First, you need a 50m or 100m straight length to find out your pace measurement accurately.

1. Set up the 100m walk e.g driveway, running track/sports centre, a known length e.g. the boundary of your land.

2. Walk with an easy, relaxed stride, counting each step as you go.

3. Repeat a few times to get an average of how many steps you take over a 100m distance.

4. Get a calculator and divide 100 (m) by the number of steps you took.

   For me 100 (m) ÷ 120 (steps) = 0.8

5. Now you have the length of your pace in metres. → [m/pace]

6. If you know your times tables it's easy to convert this to metres on the spot, anytime and anywhere.

   If I walk 10 steps, that = 8m

   If I walk 5 steps, that = 4m

   If I walk 48 steps.....that equals: →

$$\begin{array}{r} 48.0 \\ \times \quad 0.8 \\ \hline 6.4 \\ 32.0 \\ \hline 38.4\ \text{m} \end{array}$$

Gold star once you have got the hang of this

Scaling Up

Making your A4 map into a bigger copy is called scaling up. This is very, very, very useful because in effect it magnifies the site ~ helping perception to be clearer.

N.B. it is vital to keep your big map to an accurate scale, just photocopying it bigger will not give you this. The easiest way I find to scale up a map is this:

1. Outline an A4 map copy e.g. your Ordinance Survey version, with a black felt tip marker.
2. Blue tack it to a window.
3. Blue tack a big piece of paper over it with the little map fairly central (you'll be able to see the little map very clearly).
4. Make a decision how many times bigger you can fit onto the big paper – an A1 flip chart sheet usually suits a 3 x scaling up from an A4 map.
5. Then select a centre point. Make a dot in pencil. This is your anchor point.
6. Then carefully with a pencil and a long ruler measure set points from the centre dot to the A4 boundary and times this by 3 (or whatever scale you chose to enlarge by).
7. Draw a pencil line with the ruler to the scaled up points. Continue until you have enough to join the dots to create the new scaled up map.

You'll be able to see if you've done it accurately quite easily. You then can repeat the process to map existing features within the boundary onto the new map.

(A1) Big Paper

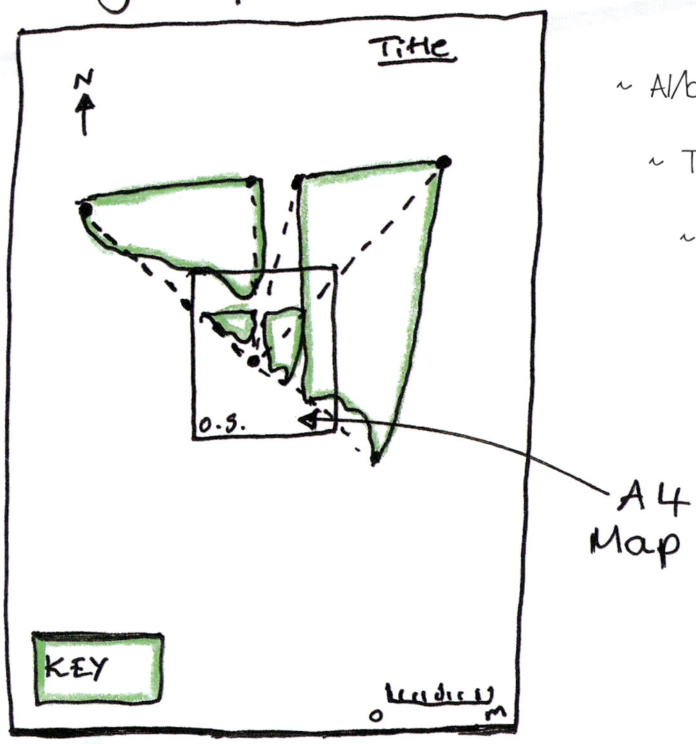

~ A1/big paper

~ The same rules apply for the big map

~ Scale, N, title, date, key

I would then make big copies to play around with. Using a dark pen to outline and then tracing the big map onto the other big bits of paper, without the need to scale up again.

A Note on Scale

Maps are often labelled with their scale expressed as 1:100 or 1:2500 etc. This is a shorter way of saying "1cm on the map = 100cm in the real world" or "1cm on the map = 2500 cm in the real world". It also applies to any other unit of measurement so at 1:100, one handspan on the map = 100 handspans in the real world.

When you enlarge a map, as described above, the scale of the new map is different to the original. Let's say the original scale was 1:100, so 1cm on the map = 100cm in the world (or 1m in the world if you prefer). If we decide to make a version enlarged by 5 times, we multiply each measurement from the old map by 5 to create our new map. This means that:

1cm on our old map = 5cm on our new map = 100cm in the world

In which case, 1cm on our new map now = 20cm in the world

So, the scale of our new map is 1:20.

If in doubt, go back to
how many cm on the map = how many cm in the real world.

Learning the Slopes

If you have a large, complex site or even a small one that contains slope(s), you will find the actual area available is more than on the OS maps or your property deeds show. This is because from overhead, the 'plan view', is a smaller footprint than on the ground.

Designing my forest garden I was totally foxed. I knew my outline map was accurate and that my pacing of measurements was too. However, whenever I paced out the distance of the elements – mainly the fruit trees I was planning to plant – there was always a discrepancy in distance. I knew my site was steep. The contour lines on the OS map showed a 20m rise in elevation/height over a 50m distance. Then one day some school maths triangle image kept looming in my mind and I realised the issue. The land area on the ground was nearly 10% bigger than when measured 'plan' view (i.e. from directly overhead).

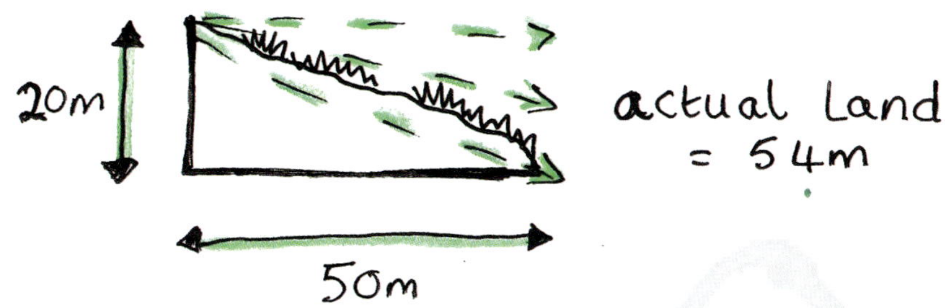

The 'Earth Care Manual' gives good info on making simple slope measuring devices, such as an A frame which you can 'walk' like legs all over your site to plot contour lines.

For my smallholding, the OS contour lines, plus lots of walking observation was sufficient to map the slope for gardens, windbreaks, fencing and so on. For our big house, planned on a very steep and irregular slope, my husband had to get serious with laser levels, bunyips and more.

Have a think and select which level of accuracy any slope on your site requires. Rolling yourself or setting a full glass of wine on the ground is informative to assess a slope too.

- ☐ I have no slope
- ☐ The contour lines on the OS map are enough
- ☐ Some measuring would help
- ☐ There are elements that need detailed surveying

Designs involving a slope for large pond systems and buildings will probably need a good, accurate map of the 'profile' (a cross section of the slope).

Sketch your impression of the slope across your whole site or of a specific area e.g. a back garden that you may wish to terrace.

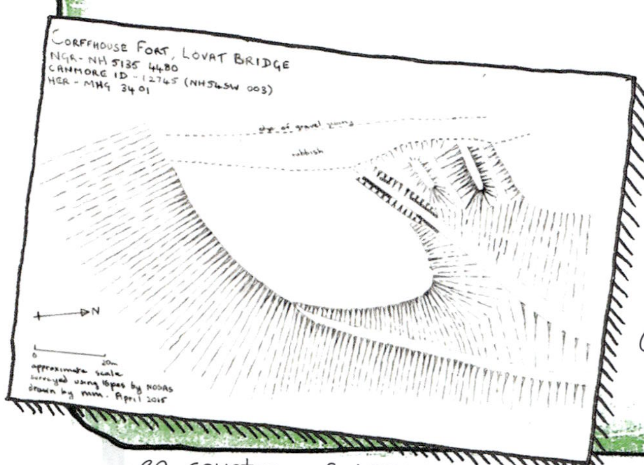

eg. courtesy of NOSAS

There are lots of different ways to represent the slope on your map.

Choose a way which makes sense to you that shows the direction of slope and allows you to distinguish between steeper and shallower slopes.

Additional Resources

There will be more games with the map soon. If you are serious about mapping, there's a few other matters to mention: useful tools include:

~ Scale rulers: these have typical scales such as 1:100 marked on a ruler so that you can easily use this on a scale map without having to convert it every time.

~ A french curve; which is a bendy thing to use in conjunction with a ruler for measuring out curvy lines on your map. Though a piece of string services pretty well for this.

~ A long (50m) tape measure or a metre wheel.

Google Earth and many mobile phone Apps have a number of GIS mapping tools and applications for profiles, slopes and so on. However, a couple of my students found these to not always be accurate once translated to actual distances on the ground, so always double check manually if in doubt.

For now there is one more SUPER resource to go with your map – MODELS.

Whether you are making a duck house, a patio garden, a mansion or designing for a big site, such as a woodland, a 3D model is fun and a super informative method of mapping your site. It illustrates information from your survey and is a design tool.

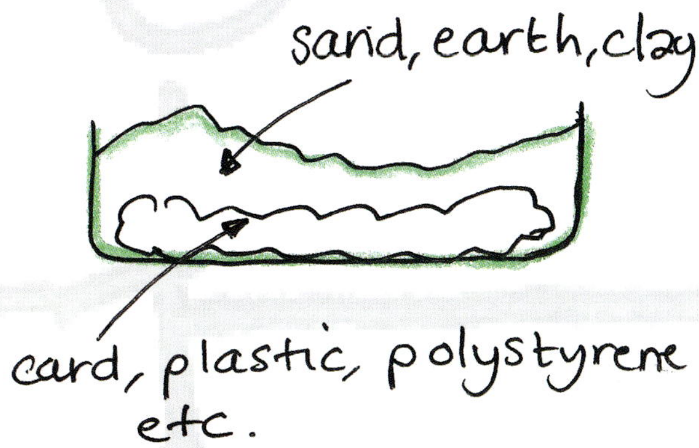

You will need some sort of tray or base of a suitable size and robust enough to move if it is getting in the way. Importantly, yes, it is ideal for it to be to scale.

Then using sand, earth, anything you can sculpt with, set to accurately modelling the landform, slope etc. Over time as you observe and walk the site, subtle dips, slopes, and so on reveal and then you can sculpt your model accordingly. Then you can add other significant features, such as trees, ponds, structures and gates.

When we are fully in the design stage of SADIMET a model makes it very simple to see the relationships between proposed ideas and their suitability from a holistic perspective. More of this later...

photo of your model.

Go on give it a go!

Base Map Checklist

- [ ] I have an accurate A4 map
- [ ] I have made copies of it for sketching my design ideas
- [ ] It includes title, date, scale, N, existing features and a key to identify them
- [ ] I have a good method for mapping details and elements within the boundary
- [ ] I know how to use my pace to measure distance
- [ ] I have scaled up my map on a big piece of paper
- [ ] I understand slopes may need a separate map and more detailed survey
- [ ] I'm making a 3D model of my landform and of the existing features of my place

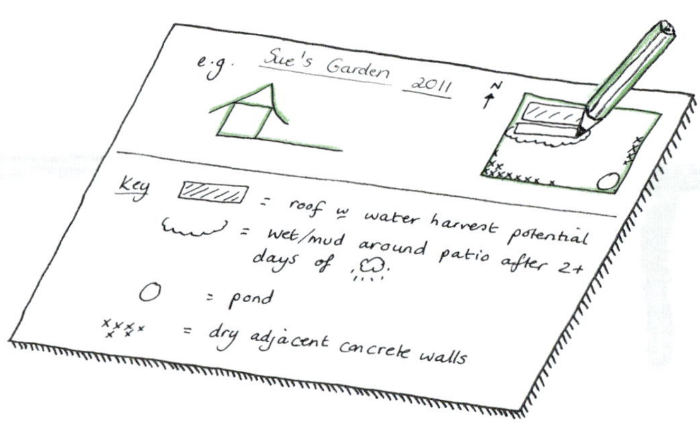

If you find making the map daunting or confusing there are plenty of resources out there from co-opting logical friends to the afore mentioned books and of course the internet

Our final job here is to troubleshoot any difficulties with mapping so that it doesn't become a suppressed worry or a blatant block – because frankly the Base Map is essential to permaculture design.

NAME UP TO 3 CHALLENGES YOU ARE FINDING WITH MAPPING & NAME 3 STRATEGIES TO ADDRESS EACH

e.g.

I don't have a long tape measure →
- borrow one off a local building site
- find your pace length instead
- sleep on it

CHALLENGES AND SOLUTIONS

1.

2.

3.

THANK YOU!

People Survey

Now we have a good sense of the land or place we are designing for. Next we need to SURVEY the CLIENT. This means a thorough survey of yourself and/or the person/people the design is for and will involve. Don't skimp on this stage, as permaculture's magic is its systems approach and the people involved in a place are as much an ecosystem as the landscape. We are not separate from nature. As with our vision statements, the views and preferences of others involved, whether complementary, antagonistic or somewhere in between, will have a large influence on whether a design is appropriate and effective.

A designer always has the challenge of extracting and interpreting the view or vision of the people involved and often this means perceiving information that the client (whether ourselves or others) may not be aware of. For example, someone involved may say contradictory things, such as wish for elements incompatible with the resources available or limited by the character of the place. The mission of this chapter is to DISCERN as deeply as we can the heart and potential of the people vision through surveying their resources, ideas, passions and situation.

NOTE: really consider WHY you want a particular element in your design, rather than a list of things you think you want. Consider carefully WHAT does a specific element do for you or the design, what is its function. For example, I'd like a mezzanine art space. When I contemplate why, it becomes more flexible and my wish is more accurately expressed as a light, clear area adjacent the living room so I can work and be close to the family. This creates spaciousness in the design and frees me from a potentially onerous commitment to making a mezzanine.

I'm going to present two styles of surveying people – the first method (Elements Survey) is a whole system approach; the second is more traditional and linear. Both are very useful, and together they are more than the sum of their parts. Fill in the people surveys for yourself and then make duplicate blank copies and repeat for any other significant people involved e.g. partner, family, user group, funding bodies and site owners.

A note on the Elements Survey: this is a change of pace and style to all the mapping and surveying tools. Head off into the inner landscape now; use the prompts under each element on the opposite page; note your responses overleaf.

Again, cast your mind wide under the headings to catch and store as much info as possible; your brain will begin to analyse and compute this with relative ease.

Use colour, doodles, stick men....

Elements Survey

EARTH

skills available

money/finance available

your own background and roots

ancestry of the place

physical resources on/off site
e.g. stone, wood, soil, stuff

WATER

areas of your life in transition e.g. kids

aspects of the place in transition or that you'd like to transform

client or other people involved in transition

areas of your life you'd like to transform e.g. get fit

SPIRIT

name 3 characteristics that are unique to you

e.g. generous, lateral thinker, practical

AIR

ideas you have

ideas influencing the design from other parties e.g. urban regeneration, school curricula, recreating a memory from youth

philosophies or ideologies

books, films, projects that have inspired your design brief

FIRE

passions

impulses

creative interests

aspects of the project you are particularly drawn to

areas of potential burn out or hot spots

The survey responses give you grounded and real pointers of where this design needs to go to succeed.

EARTH: Skills & Resources

AIR: Thoughts & Ideas

FIRE: Passion & Enthusiasm

WATER: Transitions & Flow

SPIRIT: Unique Qualities

Refer back to this survey regularly during the coming steps to keep yourself on track. You'll have a pleasant surprise now and again when you see a neglected idea, passion or such that gives you a perfect design solution. For example, a client's love of dogs, vegan art and willow sculpture once revealed the perfect combination for hiding an ugly old tank by weaving a huge puppy around it.

It's surprisingly easy to wander off into what we feel we should do, rather than what our heart truly desires, when we are designing. I humbly promise you, the latter is a sounder basis to work from and a worthwhile discipline that extends into every aspect of life. Check in with your heart's desire now. Write here a statement of intent and promise yourself to be honest with what you do and don't like.

My vision of life is

Now write what you truly wish, as if no-one else will know and as if nothing can stop you.

my true vision of life is

We only get this one wild and precious life, so if deep down we dread growing veg and truly love flowers and dancing – name it and it will surely find creative expression in your design. You will end up with an emergent solution that can take account of the 'shoulds' too, such as growing edible ornamentals (rather than potatoes and carrots) and making a dance space that doubles as an outdoor eating zone. The result will be a lot more productive and yielding. You may even inspire others to join in and grow the standard veg and end up swapping dance sessions for tomatoes.

colour in
your heart here
(a pleasant colour please!)
and bless its gentle and
fierce wishes

Traditional Client Survey

GENERAL PREFERENCES
Likes
Dislikes
Absolute no-no's
LIMITING FACTORS
Time available to create design e.g. 1 hour per week over 3 months
Time available to implement design
Money available to create and implement design

Finance desired to implement design

Ideas for finance sources

Funding constraints/objectives

Other official permissions required e.g. clearly outline and distil aims of any authority you are under and align these alongside your aims e.g. planning, site owners etc

DESIRED PRODUCE/YIELDS

Craft ~ income ~ social ~ educational ~ research ~ food (quantity and variety) ~ other

Alright, now we go into the deep my friend – our observation, vision and surveys are well established. Next we cross the border into analysis and embark on making design choices.

PART 3:
Weaving the Web

Introduction to Analysis

Analysis, or thinking tools, simplify complex factors and help us make sensible decisions. So far, we have established that tuning into our senses, observing, mapping and collecting data are tools to lay a good foundation for a design. To begin to digest all that information and transform it into material decisions, we go back to the toolkit and select a new set of tools.

As designers, the next step in the process is to weave a web from the raw matter we have collected. To achieve that we can select specific analysis tools to achieve specific ends. Imagine a carpenter, when they want to do a job they choose a tool, i.e, to make a cut they choose a saw. It would be pretty inefficient, frustrating and even hazardous to pick up a hammer and attack the task with no mental organisation as to what they were doing and why.

Analysis can be defined as the process or method of studying the nature of something or of determining its essential features and their relations. The word analysis originates from the early latin, to loosen. So permaculture tools loosen information to be formulated coherently, they are frameworks and processes for thinking about a problem or plan. They organise our thoughts and allow our mind to bring forward the relevant information or activity. Assisting the brain to weave a web of beneficial relations between all the elements in our design. If we were completely in harmony with, and steeped in our local landscape, it's possible that thinking tools would be obsolete. In the contemporary world though, we are less likely to have inherited such skills from our grandparents and be fluent in local landscape knowledge.

Analysis tools can be applied to any type of conundrum at any scale. Edward de Bono's work is an excellent source of learning and inspiration in the field of thinking and its merits.

All the analysis and design tools presented in Part 3 and 4 can be applied in many diverse ways and in many different sequences, as their quality is more holographic and cyclical than hierarchical and linear. For clarity, I am presenting them in a certain order based on my experience and introducing them one at a time. Follow this sequence in the book while you get acquainted with the purpose and outcomes of the analysis and design tools.

When you are at ease with the ideas, pick up a tool and if it works, great, if it feels clumsy and you are not harvesting any yields, then swap it for a different tool. The more you use them and become familiar with the methods, a fluency will emerge and you'll find it quite natural to move between them. I have included a brief summary and comparison of all the methods and tools this book covers in the Additional Resources in the back of the book.

Key Planning Tools

As the name suggests, this section is about tools (methods of achieving a result) in planning (method of making coherent and effective) that is absolutely key (unlocks the potential of landscape) to match your ideas and vision whilst reducing and replacing the need for energy inputs. Here we really meet the essence and magic of permaculture, albeit with a fairly dull name: The Key Planning Tools (KPT).

The KPT are a permeable interface between survey and design as the information they yield are a) part of our ongoing survey process and getting to know your place better and b) starting to design: the actual placement of elements through landscape analysis. The constraint here is that some of this information may take many years to collect and really understand, for example, the effect of wind patterns. However, we can do our best and achieve 'good enough' results by applying the KPT of sector, elevation, zone and network analysis. Fear not! What these terms mean will become clear over the following pages.

.Knowing something in advance about the landscape gets us more in the position of our ancestors and others who learn all this stuff from birth and use that knowledge fluently to maintain balance in meeting their needs from the landscape around them.

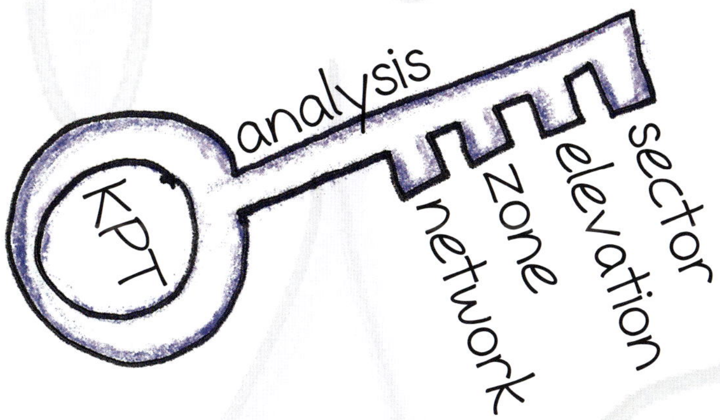

Sector Analysis

The sectors are easy to remember as they are the same as the five elements: earth, air, fire, water and wildness (spirit), plus the addition of our human perspective of the aesthetic, the view sector. Often the sectors are patterns much bigger than the site, originating from elsewhere. As with the survey data, collect your observations over time whenever possible, as the sector influence can be complex and vary immensely over time and season. Suspend hard and fast decisions, whilst allowing your observations to reveal good design choices. Use all your senses with each sector and record more formal and researched data that's relevant too. Let's take each sector in turn.

Soil Sector: Earth

This includes soil types and underlying bedrock. If you are planning any buildings or ground works, knowing the type and quantity of the rock and stones beneath you is an excellent resource. If you are going to grow food, any other plants or trees, having an idea of the soil depth, texture and type gives you a head start. If you don't match what you are growing to the soil sector you'll get feedback within a season or two that will probably make you tweak what you are doing.

I was once employed to plant an orchard for a school and found the spade met rubble and there was only 2 inches of soil, the next minute a water pipe was punctured. Analysing the soil sector on the simplest level (sticking a spade in the ground) would have shown me this. Many back gardens in newish estates are sited on rubble, so remember to check what lies beneath the ground. A few trees planted on too thin a soil is livable with, as is making a shed on waterlogged ground. Whereas designing a whole house or planting a woodland on an inappropriate soil sector can be very expensive and frustrating and may even lead a project to fail and create a lot of stress. This is a reminder to use small and slow solutions whilst we get to know our place.

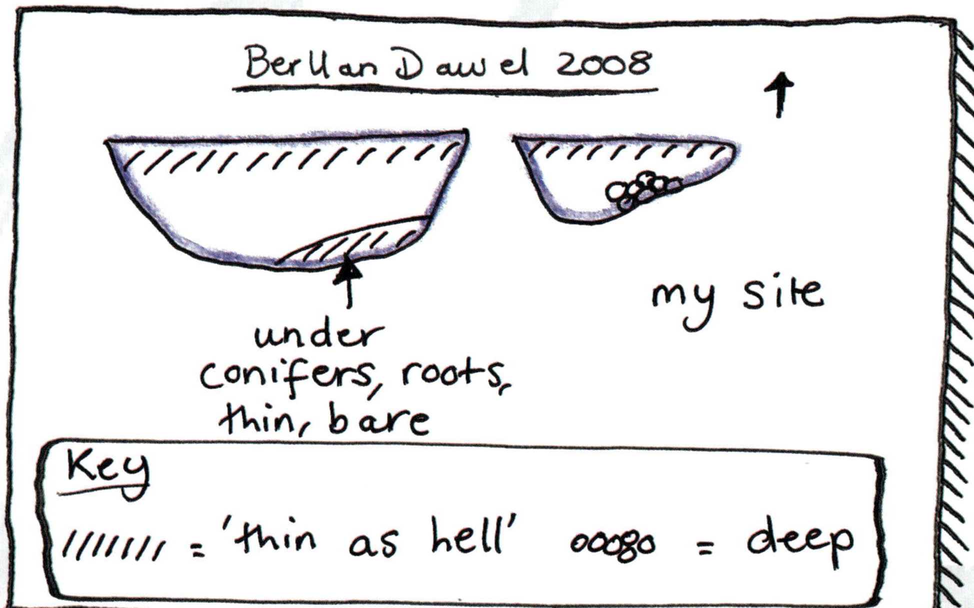

There can be great variations of rock and soil, even on a relatively small site. Best practice is to make no assumptions. My top soil is free draining shale, my neighbours a few 100m down the hill have heavy clay. A small garden can have very compacted soil where a fence or patio has been constructed, yet be perfectly friable elsewhere. It will save time and money and greatly influence the success of what you plant or build by being aware of such soil sector factors.

1. Draw a rough map of your site here

2. Sketch variations in earth/soil/stone sectors

3. Make a simple key for at a glance understanding

SOIL SECTOR OBSERVATIONS	
Soil flavour: sour, sweet, salty, bitty, alive, dead	
pH or other soil analysis	
Bedrock (online or Geological Societies)	
Contamination	
Stick a spade in and cut out a slice, draw the profile of topsoil, subsoil, root penetration etc.	
Stick a spade in noting resistance, stickiness etc.	
Walk around, note variations underfoot ~ after rain ~ after lots of rain ~ during very dry periods	
Dig a test hole, especially for any buildings or ponds, to see how fast water drains away	
Look for signs of erosion	
Official surveys data	
Research prior use (eg cattle, toddlers)	
General impression	
Other	

SOIL SECTOR ANALYSIS

This is a pivotal step in identifying the potential in the landscape and creating a place sensitive design. Consider the most appropriate locations for the elements of your design to minimise the need for energy inputs.

Combine your soil sector observations with your design vision by using the following thinking tool. Name as many things as you can in each box.

Strengths	e.g. the soil is deep from the patio to the hedge, so would suit my plans to grow vegetables
Weaknesses	e.g. the soil is thin and stony on the east boundary; poor for fruit trees
Opportunities	e.g. I could include more wild plants and herbs where the soil is poor
Constraints	e.g. manure would be good, access to the garden is too narrow for a trailer

Working conclusions based on the above..... good ideas to remember...

Wind Sector: Air

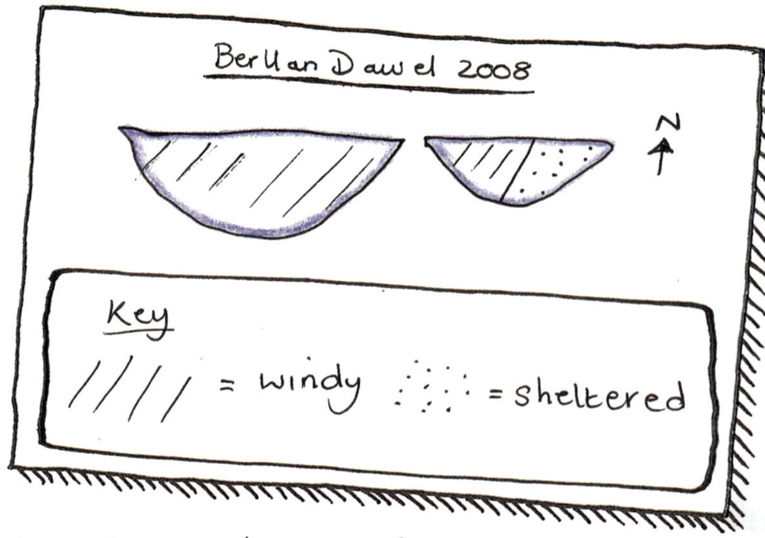

Wind is a curious creature, sometimes full of power and destruction, other times a gentle caress. Out here in West Wales it's so wet I often think we'd all biodegrade and go very mouldy without plenty of wind.

How wind moves around the globe and through our local area is complex to model. Many permaculture books have intelligent diagrams depicting landform, trees, wind turbulence and so on, giving a sense of this complexity.

However, for most purposes we need to get a good sense of where is sheltered and where is exposed and what wind does on our site when it meets resistance (e.g. walls), when it is filtered (e.g. through hedges) or what it does when it has free passage (exposed areas). Using our body as a windometer is very informative to gauge this.

As plants and houses lose up to 30% of their energy through wind exposure and as we want to obtain a yield and minimise energy inputs, the wind can have a massive impact on our experience of the 'success' of our design. Equally, getting enough air flow is essential in buildings for good air quality and for many plants to minimise mildews and cankers.

Observations of wind are crucial to design micro climates and shelter belts, whether designing houses, gardens or whole landscapes. A useful titbit is that a windbreak provides a shelter zone for about 3 times its height; walk around your site or a nearby park and test this.

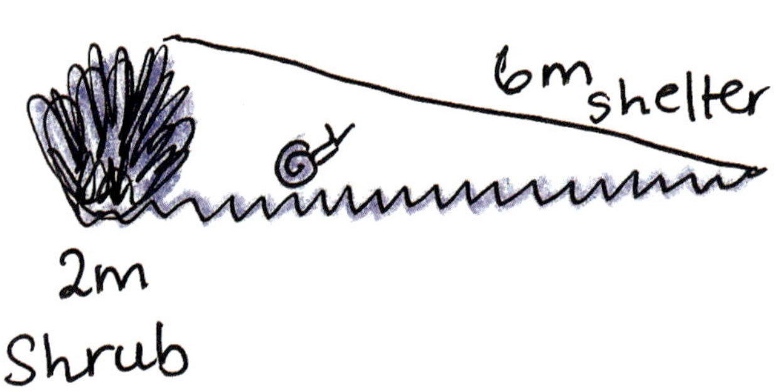

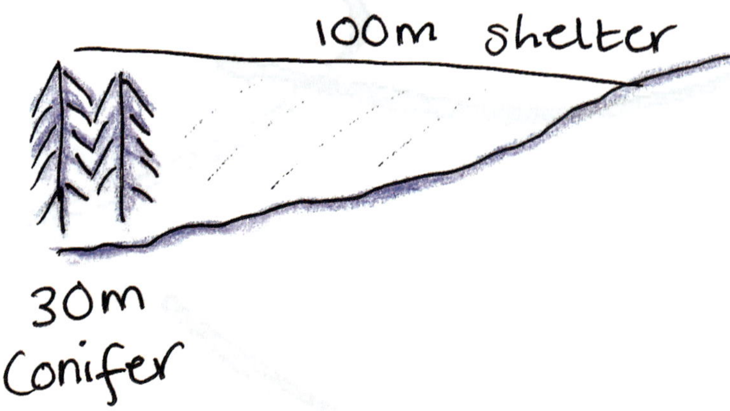

1. Draw a rough map of your place

2. Sketch your impression of variations in wind and air flows

3. Make a simple key

4. Make a sketch of how wind rises or interacts with any slope

WIND SECTOR OBSERVATIONS

use your body, ask local people and look up data to discover:

Exposed areas	
Variations of exposure at ~ different heights ~ in high winds ~ in mild wind ~ on a still day	
Predominant wind direction	
Seasonal variations ~ winter ~ spring ~ summer ~ autumn	
Variations ~ within the site ~ nearby places (e.g. the street, next door) ~ why difference if any?	
Along site boundaries ~ N ~ E ~ S ~ W	
Look for windward growing plants and trees	
Note the effects of: ~ wind by walls ~ wind by buildings ~ wind by hedges ~ wind under trees	
Other	

WIND SECTOR ANALYSIS

This is a pivotal step in identifying the potential in the landscape and creating a place sensitive design. Consider the most appropriate locations for the elements of your design to minimise the need for energy inputs.

Combine your wind sector observations with your design vision by using the following thinking tool. Name as many things as you can in each box.

Strengths	e.g. it is sheltered enough for fruit trees or veg on most of the site
Weaknesses	e.g. very windy by house and tunnels into the garden, no good for fruit trees
Opportunities	e.g. could use quick growing trees to create shelter and privacy
Constraints	e.g. windbreak takes time to grow; consider interim solution

Working conclusions... good ideas to remember...

Water Sector

As with the other sectors, it is worth investigating macro or wide-scale effects, such as flood plains and regional weather rainfall data. If your site has a river or stream, it is useful to get to know how far the water sector extends from its edges, variations in flow, volume and so on. On the local or micro scale, for those designing a small back garden or similar, the water sector can be just as relevant and significant as on a farm scale site. On the latter, understanding the water sector of rivers and boggy fields will be vital. The equivalent in the small design might be understanding about the water run off areas adjacent fences, concrete walls and hard standing (such as pavements, patios and under sheds).

Referring back to our soil sector analysis, any variations in drainage, dryness, muddy puddle accumulation etc. is worth recording here too as we are now building up more complexity and understanding as to how natural systems at our place interact.

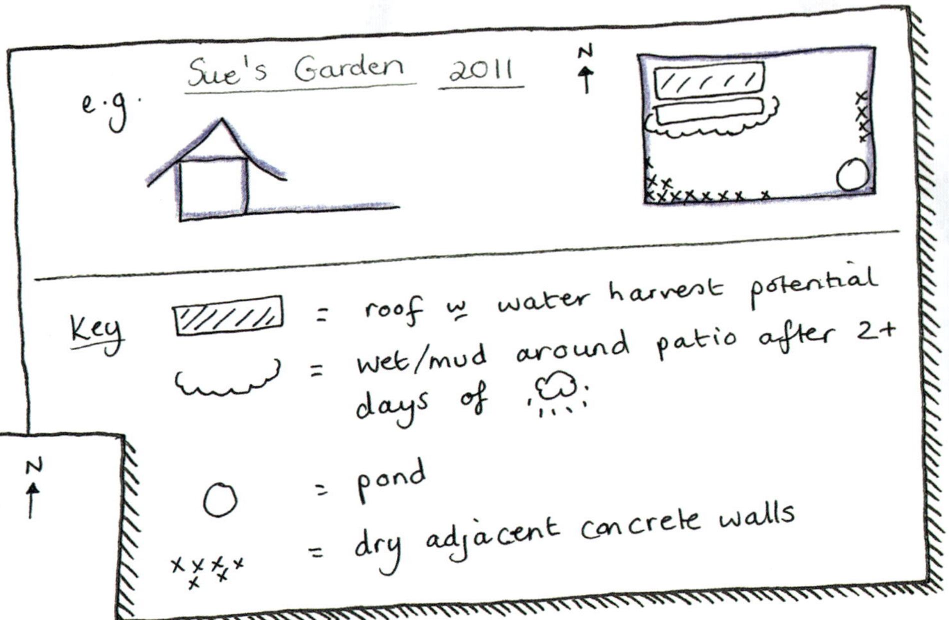

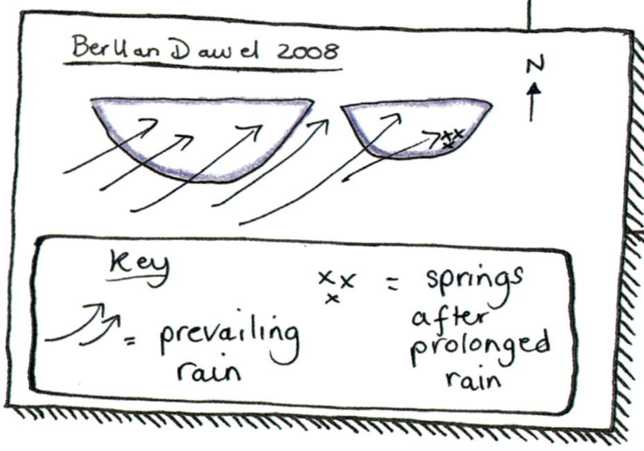

Potential for roof water collection (multiply roof area by annual rainfall), drains and existing ponds are all recorded in this section. Remember, using this tool to plan, means your observations inform good design choices by matching what's happening already in natural cycles and flows with what you'd like to see happen. For example, an area of water run off or a boggy area may well suit a pond; soil waterlogged from adjacent compaction and prior use may well benefit as a place to plant trees. In conjunction with observations of the wind sector it may reveal a wind break would suit that location too.

Remember, we are analysing through our observations and not deciding exactly what goes where. An open mind increases possibilities, whilst a decision is soon to be appropriate it also equals the end of choice (from the latin to cut away other possibilities).

A small intervention, such as a ditch or making hard standing for a gate can have big effects 100m away when we are dealing with water. So, let's be thoughtful and observe, analyse and sketch our water sector.

You know what's next!

1. Sketch a map of your site and the surrounds

2. Sketch the effects of any water sectors

3. Make a simple key to explain your drawing

WATER SECTOR OBSERVATIONS	
Predominant rain direction	
Patterns of wet/dry weather in local region	
Existing ponds and edges	
Streams/rivers/marsh	
Roofs/ downpipes and any current or potential for water collection	
Describe wet/dry zones around walls, gates, fences	
Where does water run off or drain to around hard standing? eg patios, driveways, sheds	
General variations on site	
Waterlogged or boggy ground	
General impression of drainage	
Dry areas	
Potential water effects from nearby e.g. neighbours creating ponds, trenches etc.	

WATER SECTOR ANALYSIS

This is a pivotal step in identifying the potential in the landscape and creating a place sensitive design. Consider the most appropriate locations for the elements of your design to minimise the need for energy inputs.

Combine your water sector observations with your design vision by using the following thinking tool. Name as many things as you can in each box.

Strengths	e.g. plenty of space to put water butts on downpipes for irrigation
Weaknesses	e.g. very boggy after prolonged rain around gateways and shed
Opportunities	e.g. small ponds would add habitat, beauty and relieve waterlogging
Constraints	e.g. sheltered spot for greenhouse is uphill of current water store

Working conclusions.... good ideas to remember...

Sun and Shade Sector: Fire

I'm sitting here on a cold, New Year's eve wondering if I am holding your attention with wading through the sectors in such a formulaic fashion. Then as if someone had pointed a massive torch in my eyes, I realise the sun is blaring through the clouds, my window and onto my face. The sun is just skirting the top of my neighbours conifer trees. At this time of year the sun is at its lowest point in the sky and with some satisfaction I am glad we mapped this sector carefully before planting our house on this spot. In other words at the sun's lowest ebb, we receive its light and warmth through our windows – 10m difference downhill and the sun would miss us at this time of year, 10m further up the hill and we would have been battered by our wind sector.

The plants don't mind, they're mostly underground or dormant until spring; yet for us people – whether it is for a sitting spot to watch our bird table from in a built up area or whether it's a house in a field – our desire for sun is all year round. Mapping and analysing this sector is vital for our thrival even in the depths of winter. It is important to note frost pockets (where frost stays through the day), especially in liminal periods, such as spring and autumn, when plants are still active above ground and can be nipped in the bud or finished off completely.

In spring and summer, when our plants and outdoor activities are most likely to be busy, we clearly wish to know where and how our sun/shade sector interacts with our place and how it will impact our design. As the sun is always on the move, recording your observations throughout the day and seasons is essential.

In fire prone areas, such as Australia or Portugal, serious permaculture books are recommended as they pay specific attention to this.

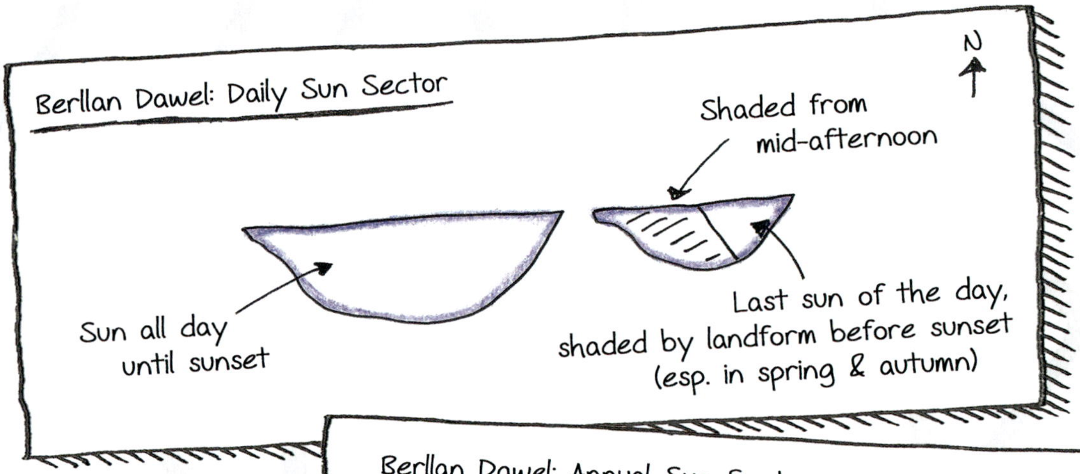

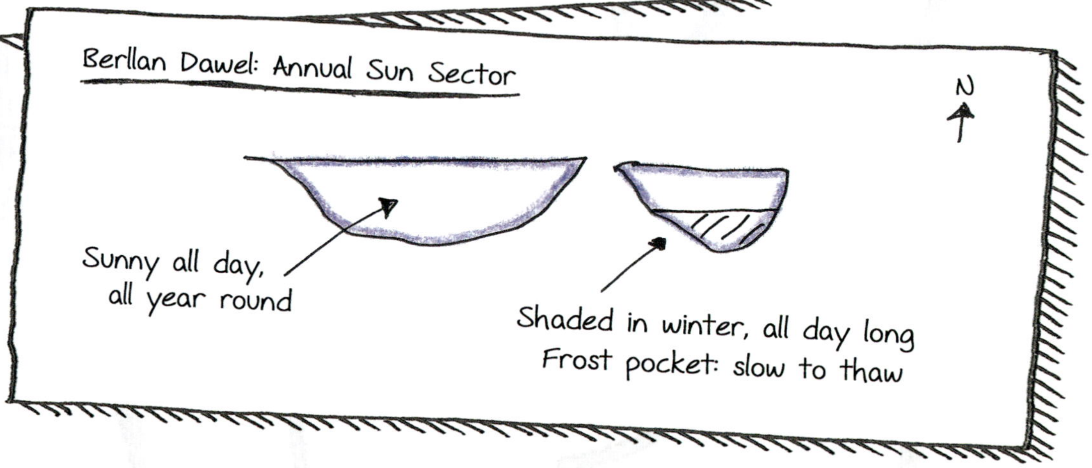

1. Sketch your base map

2. Sketch in broad areas of sun and shade

3. See if you can pinpoint the differences at times of day or season

SUN & SHADE SECTOR OBSERVATIONS

SUNNY MICROCLIMATES (areas of variation, when and where) e.g. the back door step is sunny until noon in winter 1. 2. 3. 4.	
SHADY MICROCLIMATES (variations, when and where) e.g. behind the garage, all day 1. 2. 3. 4.	
How does the sun move across your place? ~ in spring ~ in summer ~ in autumn ~ in winter	
Sun traps	
Frost pockets	
Differences in corners	
Differences in middle of site	
Next door Uphill Downhill	
Why do you think the differences occur? e.g. in relation to buildings, tall trees...	

SUN & SHADE SECTOR ANALYSIS

This is a pivotal step in identifying the potential in the landscape and creating a place sensitive design. Consider the most appropriate locations for the elements of your design to minimise the need for energy inputs.

Combine your sun & shade sector observations with your design vision by using the following thinking tool. Name as many things as you can in each box.

Strengths	e.g. sunny in morning across the whole site, suits veg and fruit trees
Weaknesses	e.g. shaded behind house, no good for herb garden and patio plans
Opportunities	e.g. small coppice would allow winter sun to reach greenhouse
Constraints	e.g. trees blocks sun in spring and autumn, casts 30m shade until noon

Working conclusions... good ideas to remember...

Wild Sector

As the name suggests this sector applies to unpredictable occurrences and energies moving across your site. I once lived at a community where I planted my first gardens, diligently applying all I could remember from my permaculture course the previous year. One day a fellow volunteer dug up all the seedlings I'd planted with no explanation. Later that day, Ken Moon, a local permaculture teacher commented on this, 'ah, that's the wild sector.' Forever after I've enjoyed the wild sector in my KPT analysis as a way of making order out of chaos as all humans seek to do, in fact need to do to thrive.

Here at the Lammas eco-village, whether neighbours, visitors or myself, people seem to be the dominant wild sector at work on the site. Film crews trampling plants, strangers looking for phone reception, or even 100m of my plot removed by a neighbour to change their drainage arrangements one day.

I've heard urban permaculturalists refer to the appearance of skips – unexpected bounties of resources – as their wild sector or wild edge. So the wild sector is up to you to interpret, in your own context, what you experience as wild – whether it's freak weather or wildlife, rabbits moving in from miles away and eating your greens, rats eating your wiring, moles in lawns or local development plans originating from miles away – I reckon it's up to you what counts as wild sector influences.

By its nature unpredictable, the wild sector may reveal sporadically and suddenly. All you can do is anticipate and then tweak your design and approach to the sector effects.

24 hours after writing the paragraph above my house burnt down to the foundations. In mere minutes it spread beyond control, a few hours later the fire crew knocked down the remaining fireproof walls of lime plastered straw and cob. Now that is a wild sector and feeds into the design of this book quite radically, as this Permaculture Design Companion is based on my 9 years living at Lammas. Following the fire, now months later, I'm having to face the likelihood of not returning to my land and my work (there at least) as a permaculture teacher. So, I'm observing myself, analysing the options and tweaking the design to simplify this book and push on with getting it finished. The effects of the wild sector for me have been a huge call to make space for thrival rather than struggle and survival.

You can only plan and design so much, wild sector reminds me to be flexible and be able to live well with uncertainty and change.

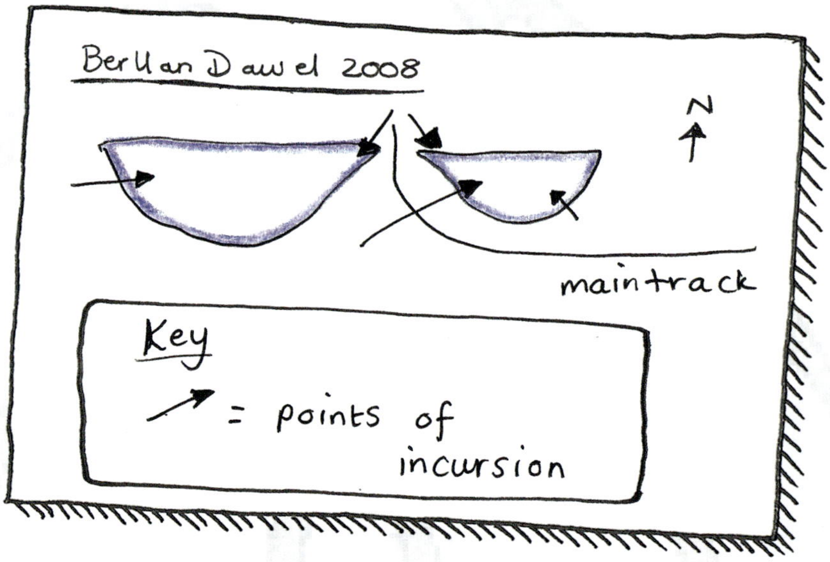

1. Sketch your site

2. Mark zones and flows of potential wildness / unpredictability

3. Sketch existing areas where wild plants and animals are active

WILD SECTOR OBSERVATIONS	
Wild creatures ~ presence ~ routes ~ frequency	
Wild plants (include invasive ones)	
Neighbours / Visitors	
Local development	
Extreme weather	
Unpredictable factors at your site or other places ~ past experience ~ present ~ future	

WILD SECTOR ANALYSIS

This is a pivotal step in identifying the potential in the landscape and creating a place sensitive design. Consider the most appropriate locations for the elements of your design to minimise the need for energy inputs.

Combine your wild sector observations with your design vision by using the following thinking tool. Name as many things as you can in each box.

Strengths	e.g. I've lived here for years, I know what to expect
Weaknesses	e.g. my neighbours are unpredictable; edible shed idea may offend
Opportunities	e.g. zone surprise visitors with shrubbery to permit nude gardening
Constraints	e.g. planning restrictions may inhibit my forest school ideas

Working conclusions... good ideas to remember...

View Sector

The View or Aesthetic Sector can be crucial to whether we enjoy being at a place or not. Whilst logistics, energy and ecological efficiency are all vital concerns to managing our resources sustainably, a good view can make the difference to our soul. If this is neglected we might find we just can't be bothered to do things well, that some fundamental part of ourselves cannot be motivated and enjoy the activity we need or wish to do and the place where we do it.

When we took on our smallholding, the blank, denuded fields, were generally uninspiring and challenging due to slope and wind. The only view appeared to be a boring block of dark conifers on our neighbours plot. We diligently continued mapping the other sectors, filling in the landscape survey and concocting various ideas for our livelihoods and planning a super efficient and ecologically robust design. I just kept pushing away the feeling of demoralisation, doing my best to appreciate what we had and the potential for creatively transforming an empty field into a biodiverse haven. Only when we fully acknowledged how disappointed we were to have only the conifer block to look at, did scope appear to work with it. By then turning our attention explicitly to the view sector, it became apparent there was a gap, a viewscape did exist. It was subtle. By noticing it and then combining with other sector analyses, we were able to optimise that slither of view and site our house to benefit it. Everything began to fall into place, both in terms of physical location of the design elements and in terms of our enthusiasm to root ourselves on that land. The knock on effect of this for the huge work that lay ahead was tangible. When people visit now, the view is often the first thing they remark on, this was created by siting structures, planting and paths to capture it.

Expensive houses have good views and popular destinations usually have exceptional views for the simple reason they make us feel better or even lift our thoughts to a more expansive and positive state. Artists are compelled to recreate and immortalise beautiful views; sacred sites, such as temples, often occupy a high point with outstanding views for their resonance with god or an elevated state of being. Centres of wealth and power, from cathedrals to international institutions, sit in the landscape with commanding views.

Hill forts, castles and country mansions look out across the landscape, carefully sited to maximise the view. Why did they do this? Usually, practical reasons of defence from attack are cited, or the superior status derived from looking down on the masses. Recently, walking around our local area on the sites of ancient forts, my husband and I could not dispel the feeling that these places were chosen as much for the beauty of the view and the astonishing effect a good outlook has on the human psyche as for any strategic or power-over motivations.

Your project may well be in an urban area, or other location where the view is restricted beyond your control. In this case, it's still important to map the view sector, even if it's all negative from your perspective. There are many ways to mitigate a poor view e.g. murals on ugly walls; trellises; trees and tricks to lead the eye elsewhere. First you have to identify where the view is and from what perspectives and distances it is significant. Then we can set about the business of employing the imagination to deal with it.

1. Sketch a map of your place

2. Show with shading or dotted lines the most significant views (good or bad)

3. Make a key to explain the view sectors

VIEW SECTOR OBSERVATIONS	
Important views to maintain ~ From the site ~ Within the site ~ Looking onto the site	
Unpleasant views ~ From the site ~ Within the site ~ Looking onto the site	
Different times of day and night that the view has significance	
Different seasons affecting the view ~ Winter ~ Spring ~ Summer ~ Autumn	
What differences in the view affect you and why? How do they make you feel?	

VIEW SECTOR ANALYSIS

This is a pivotal step in identifying the potential in the landscape and creating a place sensitive design. Consider the most appropriate locations for the elements of your design to minimise the need for energy inputs.

Combine your view sector observations with your design vision by using the following thinking tool. Name as many things as you can in each box.

Strengths	e.g. lovely view of valley from most places, suit house or BBQ idea
Weaknesses	e.g. the road dominates; not enough space to grow trees to screen it
Opportunities	e.g. living willow structure to block road and meet bird haven idea
Constraints	e.g. important existing windbreak trees block main view from living room

Working conclusions.. good ideas to remember...

Sector Summary

Phew! On a course we'd spend 45 minutes per sector. On your own place I send a wish that you will revisit sector observation and analysis MUCH MORE OFTEN than that with your unique bodyometer and continue to add to your sketches and notes in the prior exercises.

One last totally important point to consolidate this whole sector business, is to transfer the broad info for each sector onto your base map, or a dedicated copy. I generally use shading of different colours and dotted lines to depict the sectors. However, OVERLAYS are very, very, very good with sector analysis and designing. An overlay is a transparent / tracing paper layer. You can use one layer for each sector (that's lots of layers to caretake) or one for the sectors combined. When my husband designed our big house, sun, water and soil sectors were so crucial to the design he used tracing paper for each.

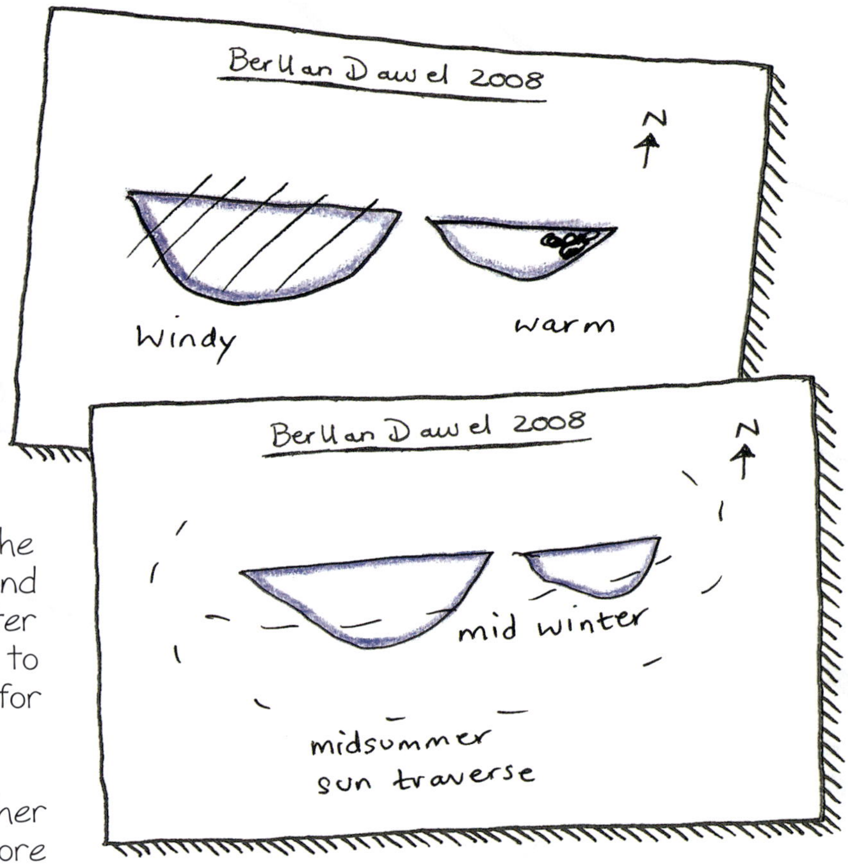

I'll leave it to you to decide whether to employ overlays or mark more simply on a map of your site. Either way – make it so!

TICK THE APPROPRIATE BOX

- ☐ I mapped my sectors on tracing paper and overlayed them on my big base map
- ☐ I decided to map the sectors clearly and simply on the actual base map
- ☐ My proposed ideas in relation to the sectors go on separate copies of the map, not to be confused with the base line data

+ POSITIVE ~ Name 3 things that made you feel good about the potential of your place or about yourself in relation to the sector analysis:

- NEGATIVE ~ Name 3 things that came up that are challenging about the site, or to you, in relation to the sector analysis:

? INTERESTING ~ Name at least one thing that caught your eye / stood out / or was unexpected in the sector chapter and process of analysing the sectors:

Evaluation

Slope, Elevation & Aspect

The degree of SLOPE at your place will definitely have a bearing on any plans that involve gravity e.g. waterflows for pond systems; or designs that require a flat surface e.g. a building or a wheelchair accessible picnic area. In wider terms, even if our design vision is relatively passive, such a small sunbathing garden where you can look at flowers, the degree of slope affects what level of exertion we need to do anything. In other words how much energy we will need. A smallholding that is quite steep has considerations like hauling 10 tons of manure around or catching and storing rain water from roofs. Many factors are greatly influenced by how steep the terrain is.

ELEVATION means how high your place is above sea level – on an OS map the contour lines tell you what the general slope is over an area. As I mentioned in the mapping chapter my site is like this:

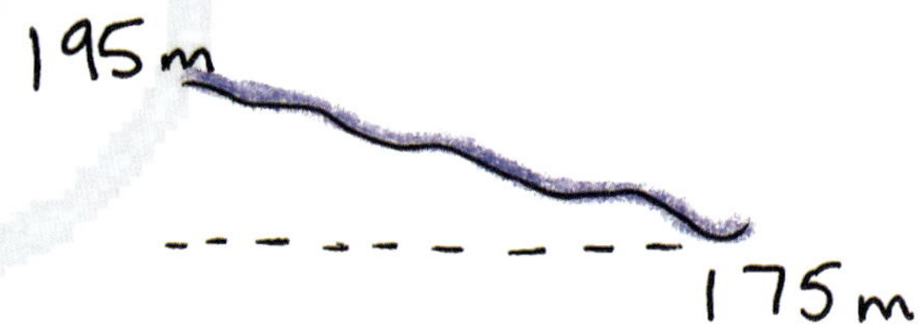

By now you will probably have a sense of whether your design requires a 'to scale' cross section of your slope profile.

The elevation also influences what you can grow as it affects temperature – a higher site is cooler and spring comes later. A sloping site, however high its elevation, is less likely to hold frost and the sun intensity will be greater, as although it is colder at a higher elevation, on a slope you are angled more acutely towards the sun.

Angle relative to the sun is known as ASPECT, in other words it's the direction your site slopes ~ a north, east, south or west aspect. Many elements benefit in the northern hemisphere from facing south towards the sun for warmth and light.

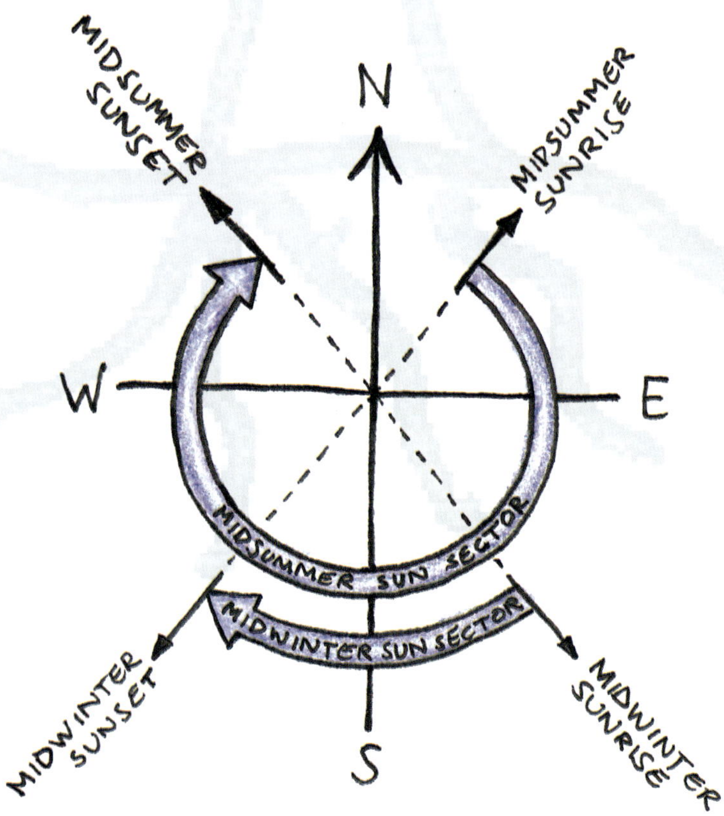

Two points to mention:

At midsummer the sun gains a whole 45° of aspect in the morning and evening. If your design is likely to be active at this time e.g. veg gardens or camping, and your sun sector is not shaded out on the NE or NW – it makes sense to design for this wider aspect rather than purely focusing on south facing areas at your place.

If your site is open and yet slopes in one direction, it's more helpful to see the available sun sector rather than focusing on the aspect. For example, if you have a north facing slope, unshaded to the north yet limited by tall trees to the south this would be the case. Similarly a flat site has no particular aspect – except 'open.'

ELEVATION EVALUATION

My site is: m above sea level

This means it is milder than higher areas and cooler than lower areas

suitable for:

unsuitable for:

(remember creating microclimates with trees and shrubs makes a place warmer and more feasible for tender plants and fruit trees)

It has a
- ☐ no slope
- ☐ gentle slope
- ☐ steep slope

The elevation at the top of the site is:

The elevation at the bottom of the site is:

ASPECT EVALUATION BOX

☐ My place faces this is its aspect.

☐ My place is flat which means it is open with no particular aspect

Do you have significant shade from trees or other buildings?

Describe:

SLOPE EVALUATION

Slope can vary a lot within a place. Sketch here one or more slope profiles (cross sections) of your place.

An area that looks flat often turns out to be sloped – roll a ball or yourself on the ground or use a level to find out. Even better, regularly traipse around carrying heavy things and then add more detail on slope to your survey, maps and model.

Our field looked uniformly sloped from the N to SW – we dutifully modelled this and kept it in mind for the placement of gardens and buildings later in the design process. However, over some months of hauling toddlers around whilst surveying, my husband noticed a significant difference in the landform than we'd noticed before. An area protruded and then dropped steeply down and faced a different direction to the rest of the field. Once transferred to the model it became a clear winner for the site of a small, earth sheltered dwelling. We had not noticed it previously as the nobble led to a steep, shady corner, which subconsciously we always avoided with the pushchair.

THE LESSON BEING: just because we avoid part of our site and don't consciously notice it, doesn't mean it's not there. It is there and can be totally intrinsic and useful to a landscape sensitive design, harnessing energy from the very placement itself.

In relation to the elevation, slope and aspect of my place:

ITS STRENGTHS ARE;
(e.g. a slope for a pond to irrigate further downhill, a suitable elevation for fruit trees, a south aspect for a chalet)

1.
2.
3.

ITS WEAKNESSES ARE:
(e.g. too steep to camp on, an area receives too little sun for vegetable growing, the sunniest place is too windy)

1.
2.
3.

IT PRESENTS OPPORTUNITIES TO
(e.g. grow microclimates to create sun traps and habitat)

1.
2.
3.

ITS CONSTRAINTS ARE:
(e.g. the main access is at the top of the site and this is...)

1.
2.
3.

Zone

A zone is an area defined by the frequency of activity that occurs there. Like all systems in real life, zones interact with each other and activities will cross zone boundaries and have relationships between them. As a zone is focused on activities and how often they occur, e.g. harvesting firewood once a year or picking salad every day, it has a great bearing on our energy efficiency. In time, doing stuff, we soon learn economic routes and methods by experience of a spectrum from flow to resistance.

As permaculture designers our aim is to minimise effort (inputs/need) and maximise gains (outputs/yields) and to discern appropriate locations for our design elements and functions that constitute our activity in a place.

The phrase 'in the zone' is used by athletes, business people and creatives of all kinds to describe the efficiency of flow when in the correct zone, whereby a state or process achieves effortlessness.

Depending on the scale and nature of your project, the activities in each zone will vary. I've given examples below to suit someone with a few feet of space up to a large scale. In permaculture, the zones are usually defined as:

zone zero zero	the self ~ constant habitation; attitude, approach, awareness
zone zero	central point, usually a building ~ all activity originates from here
zone one	daily attention e.g. a salad garden; polytunnels filled with seedlings
zone two	frequent attention e.g. a birdfeeder filled fortnightly; crops cut monthly
zone three	regular, infrequent attention e.g. a tree in a pot; a shrubbery; an orchard
zone four	once or twice a year attention e.g. trellis on a balcony garden; a coppice
zone five	rarely or never visited, e.g. a spider web; a hedge left alone to be a wildlife corridor; acres designated for rewilding.

You may decide to have some elements zoned temporarily, for the short term, whilst you set up your design or to learn more about the site until you zone the activities for the long term. For example, a small garden may be zone one for the early years, to become a much bigger zone one elsewhere after some experience. A caravan might be a zone zero whilst you build a house and later the caravan area become a zone four windbreak and firewood crop. The permaculture principles 'small and slow solutions' and 'design from patterns to details' comes to mind here. Allow space for events to unfold and flexibility for implementing your design when it comes to zoning the activities at your place.

If you are working with a model, playing with different ways of zoning gives you a great overview of the whole site. With your surveys and key planning tool observations in hand, you'll find activities begin to naturally fall into good, appropriate locations.

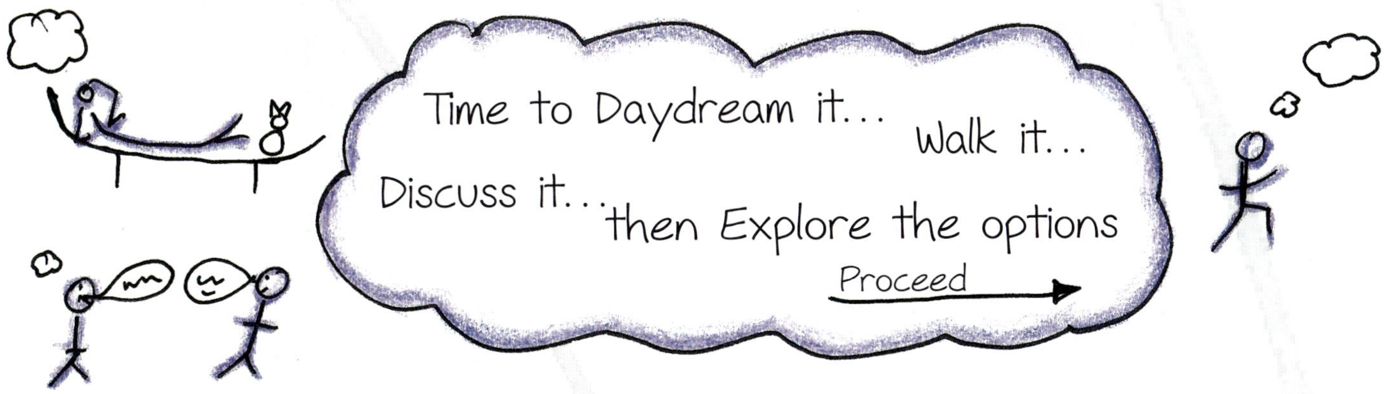

We're going to do some exercises on paper – remember, like most of the tasks we are doing, by walking the site, imagining the realities of activities and observing what works at other people's places, makes the exercises meaningful. It may be the case that your place has existing zones (pre-dating your design being settled and implemented); you'll need to bear these in mind as we analyse suitable zones for the activities in your design vision.

Let's recap our main elements. After writing your vision statement you bullet pointed up to 10 elements. Through the survey and mapping processes and the application of the key planning tools, your plans may have changed as your supercomputer sifted through their suitability in relation to your landscape.

Now is the moment to re-state them, more may have emerged by now too. Please ensure you have included one wild idea, something that pushes the edge beyond the classic elements such as herb gardens and orchards. For some reason, most my students choose a sauna or japanese bath house as their wild idea. I guess that is the result of weeks camping with a bucket shower...

MY MAIN ELEMENTS ARE:

1
2
3
4
5
6
7
8
9
10

Now, by frequency of activity and the attention required, locate each element above in its appropriate zone. Doodles and stick men welcome as they are a legitimate right brained design tool.

zone zero zero

zone zero

zone one

zone two

zone three

zone four

zone five

Network Analysis

As the name suggests, network analysis is about how all the different elements of the design are connected and interact. Test ideas on the maps and your model. Walk the routes between the elements and zones, imagining the reality of moving between activities. Walk at varying paces from super speedy to very slow; walk around with someone who has relevant experience (e.g. an experienced gardener) or who knows the situation well (e.g. a family member). Bear in mind we are aiming to establish beneficial relationships wherever we can, as therein lies the elegance and ecological and energy efficiency of the design.

Sketch your site above. Shade the zones as you currently see them working out. N.B they don't have to be circles, e.g. a zone 4 shrubbery can weave between zones, doubling as a windbreak and wildlife corridor. Draw a symbol for each significant element and doodle with a pencil the network connections and routes between the different elements. Test the zoning on your model and base map copies.

+ POSITIVE

What positive flows can you see in your current design layout?

- NEGATIVE

Where is there resistance to moving between activities?

? INTERESTING

What random or interesting ideas occur to you in relation to zone and network analysis?

Evaluation

Input <-> Output Linkages

A flourishing ecosystem links its inputs (needs) and outputs (yields), as does a society that lives in balance with its resources. A wild wood or a native tribe doesn't require inputs such as gas piped to it or to stock up at a supermarket to meet its needs. There are no outputs, such as rainwater, being wasted or plastic rubbish polluting the system. The energy of the sun and appropriate, relative location of all the different elements for life, create an energy efficient system; all supported by beneficial relationships that create resilience and support the whole. Bill Mollison made the important point that an unmet need = hard work or fossil fuel use; and an unused yield = waste or pollution. By using the unlimited potential of our creativity, we can design something that minimises waste, hard slog and fossil fuel inputs.

In our design process, linking inputs and outputs is a key tool to achieving efficiency and resilience and creating positive feedback loops. An example might be a roof. Its needs might be structural, aesthetic, spatial and so on and its yields might include keeping us dry and warm, a nesting site for birds and rainwater catchment. Then, when analysing other elements you can identify links. The roof becomes a valuable source of water for a polytunnel or bathroom. The need for the roof to have a low visual impact might suggest the roof design being turfed, which would in turn provides other yields, such as habitat for insects and so on.

Take each of your elements in turn, writing down ALL the inputs and outputs in your supplementary research book, including undesirable ones. Think as laterally as possible and fill in as much specific info as you can around quantities of materials, amounts of time, space required and so on. Here is an example to give you the idea.

ELEMENT: Christley's Yoga House

INPUTS/ NEEDS	OUTPUTS/ YIELDS
Materials (4 tons timber etc)	Spare materials/offcuts etc
Expertise (design, construction etc)	Skills accrued by self and others
Finance (£16,000)	Income possibilities (£300 per week)
Space requirements (8m x 8m)	Shade/shelter etc
Design development (more visioning)	New ideas/ potentialities
Ambience (soft, warm light)	Wellbeing/ health of self and clients
Water supply (100 litres per day)	Grey water/ roof rain water collection
Electricity supply (2kwh per day)	Lights / power for stereo
Time to build it (5 months)	2 volunteers helping and learning
Heating (3 tons wood autumn/winter)	Heat
Clients (10 x 3 days a week)	Noise/ traffic (3 x a week, 10 cars)

When you have completed the inputs and outputs thoroughly for every important element in your design, start identifying linkages between them by analysing them alongside each other.

Here is an example of how Christley made links between all her design element inputs and outputs for her smallholding and yoga centre.

ELEMENT	INPUT/NEED	MET BY	OUTPUT/YIELD	HARNESSED FOR
Buildings 	Water Power Light	Bore hole/rainwater Solar PV panels Solar PV panels	Grey water Electric	Zone 1 irrigation, domestic water, customers, surplus electric to grid
Orchards	Shelter Fertility Pest management	Hedgerow / siting Animal manures Chickens/ species	Crops/ income	Food needs Produce for sale
Horse livery 	Shelter Grazing Water	Existing barn Existing fields Roof rainwater	Income Manure Horse basic needs	OPD livelihood Crop fertility Income
Bees / honey 	Shelter for hives Fodder/nectar Storage Market	Trees, hedges etc. Comfrey etc. Existing barn Local opportunity and visiting clients	Honey Pollination Resilience	Income and local economic and ecological resilience
Strawberries, vegetables and herbs	Shelter Fertility Market	Polytunnel/ planting to establish microclimate Animal manures Local opportunity	Reliable crops Good yields Income	Low input- high output sustainable enterprise
Salves	Plants and siting Retail market	Dual function of plants as soil repair species Local area and well being/yoga clients	Compost/ fertility accumulator plants Important bee fodder Income and health	Low input- high output sustainable enterprise
Well-being clients 	Venue/ facilities Skills	Proposed well-being centre and existing barn Christley's experience	Health, well-being, income, customers for other produce	Low input- high output sustainable enterprise Asset to local area and venue for other related training services for hire
Firewood trees 	Initial planting Annual coppicing Storage	Local permaculture group and friends and family Barn	Firewood Habitat and biodiversity Shelter	Carbon neutral heating Other crops and house benefit
Waste	Grey water Organic matter	Reed beds and ponds Compost systems	Clean water Fertility and carbon sequestration	Crops and biodiversity habitat Crops and benefit wider society

Fill in the table below to record any potential linkages. Ask a friend to have a look to see if there is anything you may have missed.

ELEMENT	INPUT/NEED	MET BY	OUTPUT/YIELD	HARNESSED FOR

Elements & Functions

EVERY ELEMENT SUPPORTS MORE THAN ONE FUNCTION (=> EFFICIENCY)

ELEMENTS (e.g. a pond, a greenhouse, a kitchen) are distinct from their FUNCTIONS (storing water, growing salads, cooking). For best practice we will now make sure every element you wish to include in your design fulfils at least two functions; this is for energy and ecological efficiency. A further way to reinforce resilience and create a diverse network of beneficial interactions in your design.

For example, a woodstove (element) might support the functions of cooking, heating water and space, creating ambience and drying clothes and herbs. Another example: a firewood coppice of diverse trees (the element) might give food, fibre, habitat, windbreak, control erosion, provide privacy and fix nitrogen for soil fertility. See why ensuring that every element supports more than one function creates efficiency? I love it!

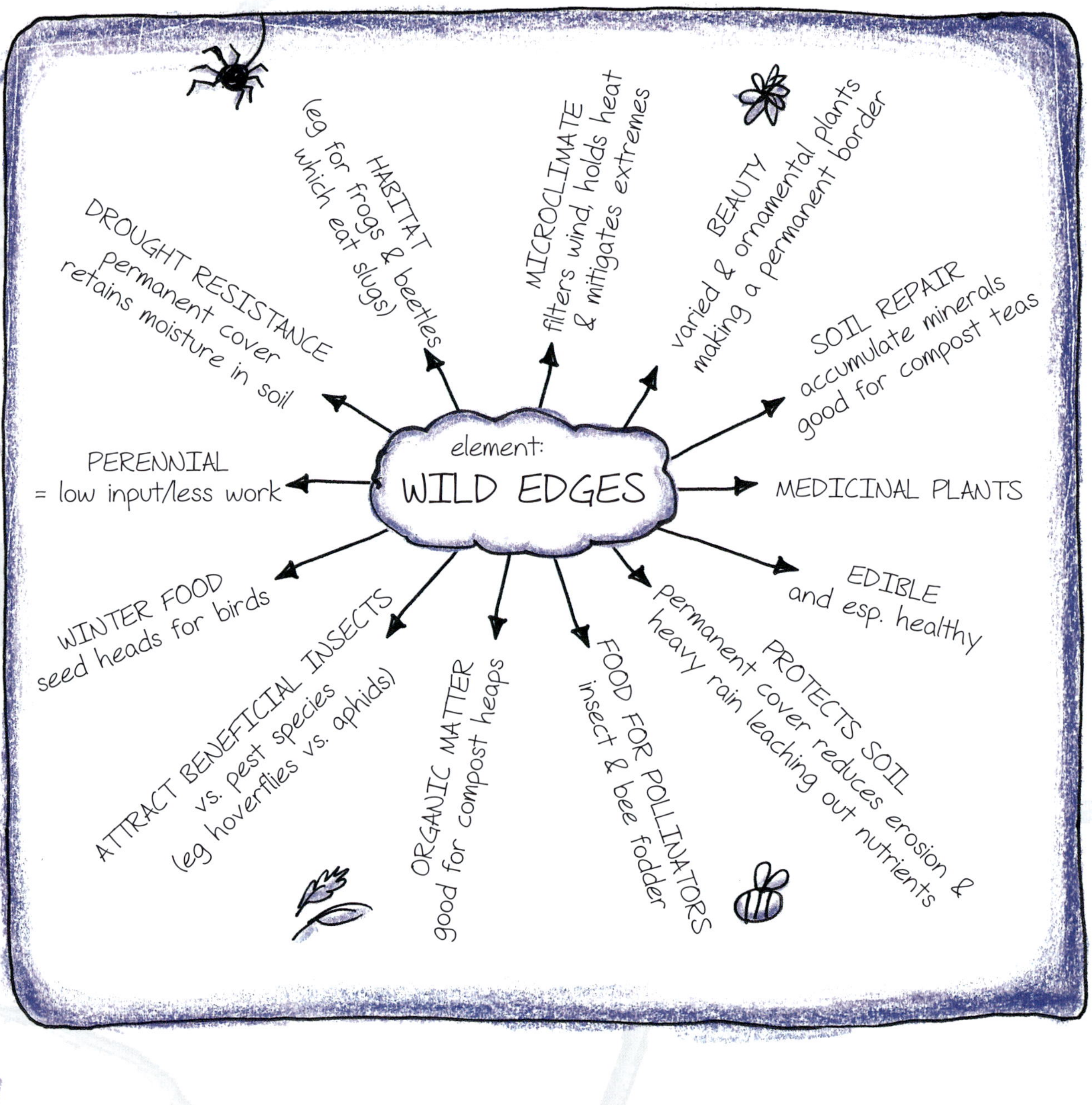

Now take each and every one of your elements in turn and map which functions it supports, seeing it on paper will prompt more ideas. Use the boxes below to get the hang of it and then complete your other elements in your folder or notebook. If you only can think of a couple of functions, consider developing the element into something that would create more efficiency e.g. an ornamental plant bed will support more functions by ensuring some species are included that bring in useful pest predator insects or have seeds that birds can enjoy in the winter.

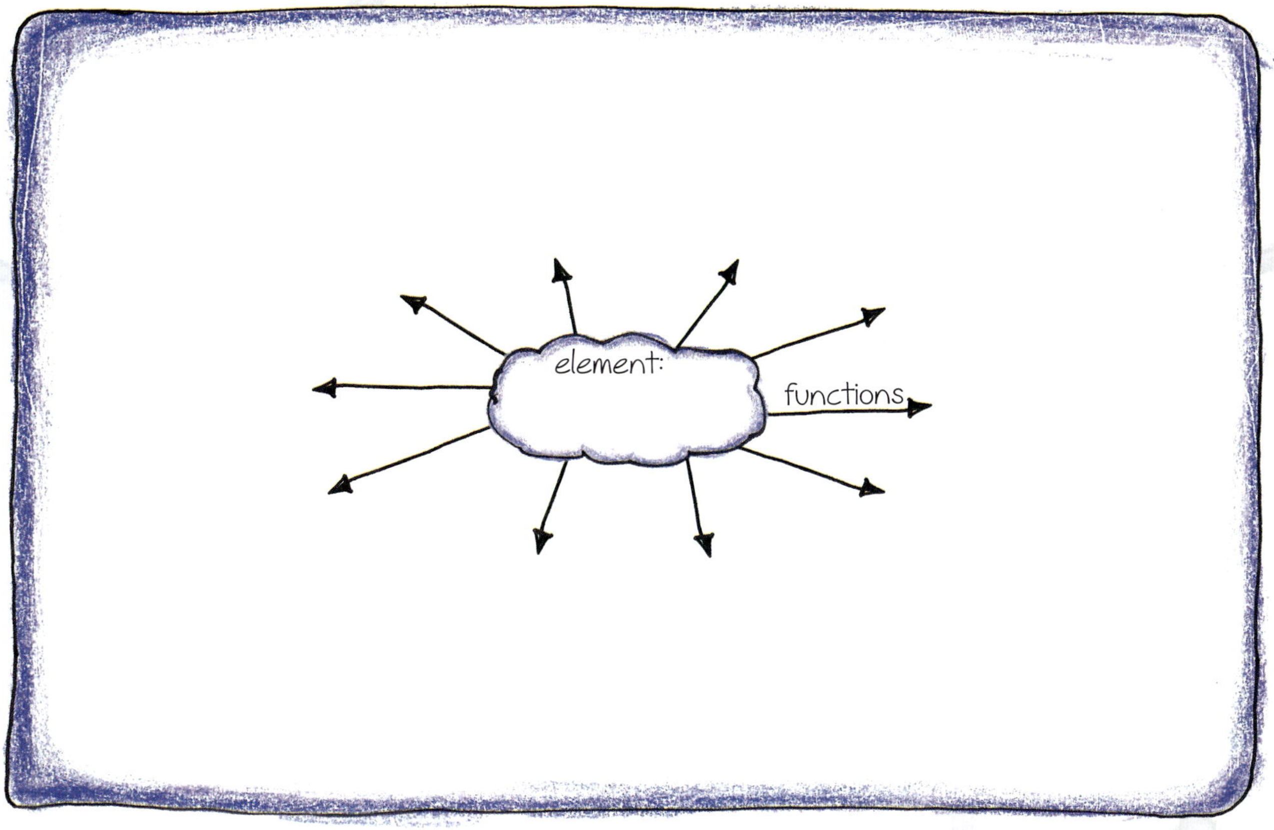

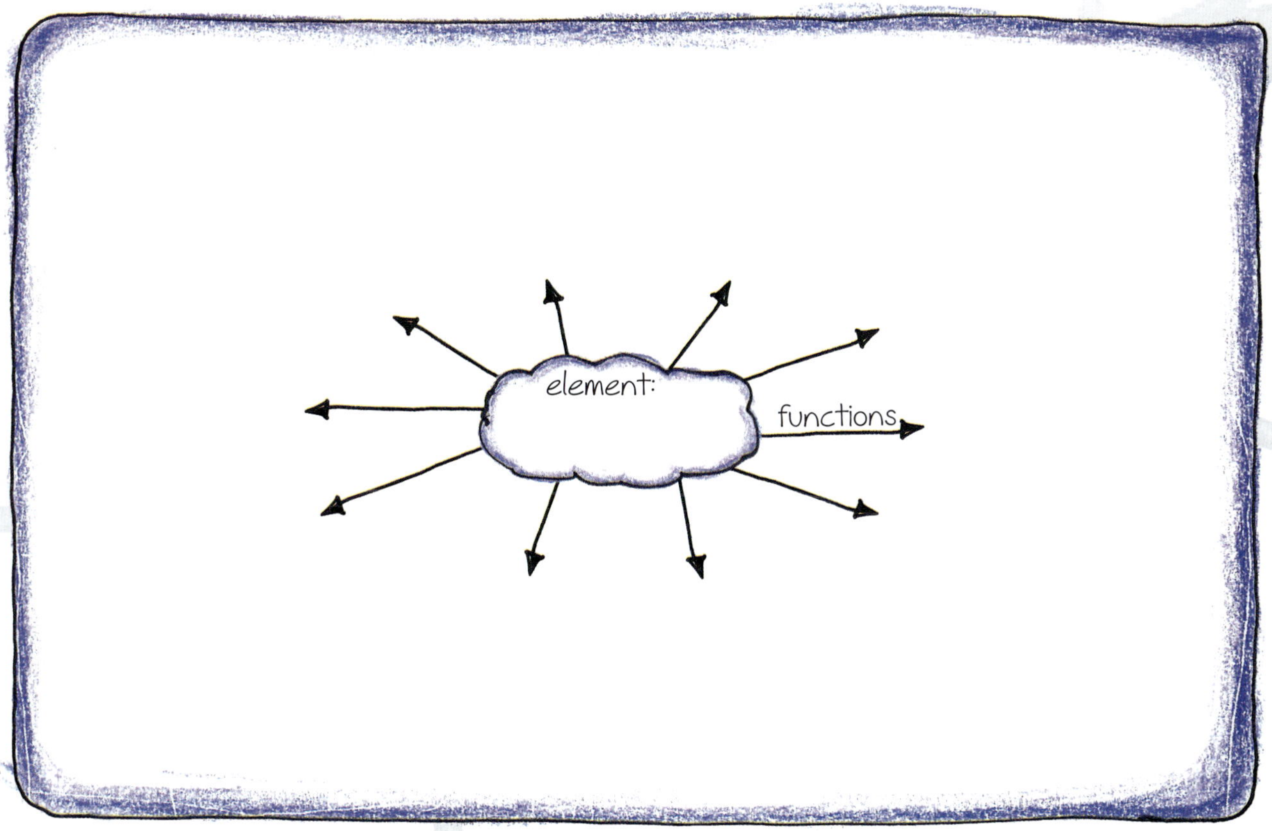

EVERY FUNCTION IS SUPPORTED BY MORE THAN ONE ELEMENT (=> RESILIENCE)

Now, ensure every function will be supported by more than one element, complementing the previous exercise. This builds in resilience in the case of one or more strategies failing from climatic, environmental, market, financial or personal circumstances. A function is an activity that is natural to, or the purpose of, a person or element. Heating, washing, eating, sleeping, playing, studying and so on are functions that have a purpose in our lives and may well be reflected in our designs. Outdoors, particularly for garden based designs, functions we are wishing to include might be provision of fertility, pest control, shelter, beauty, storage, versatility of a space and so on.

Now scan through your whole design, one area at a time, one activity at a time and really get a feel for what are the functions you wish to achieve, what is the purpose of all those juicy elements going to serve in context of the people whose vision this is for. You may find it easiest to lie down and close your eyes and really drift into the design as if you were going through scenes in a film and then note down the functions.

THE MAIN FUNCTIONS OPERATING IN MY DESIGN ARE:

1.
2.
3.
4.
5.
6.
7.
8.
9.
10.

Consider each function in your design. Work with 2 functions in the boxes below, as shown in the example. Continue the exercise in your notebook for the other functions you have identified. List as many elements as you can that will support each function. Use this exercise to develop other aspects of your design so that they too can support functions.

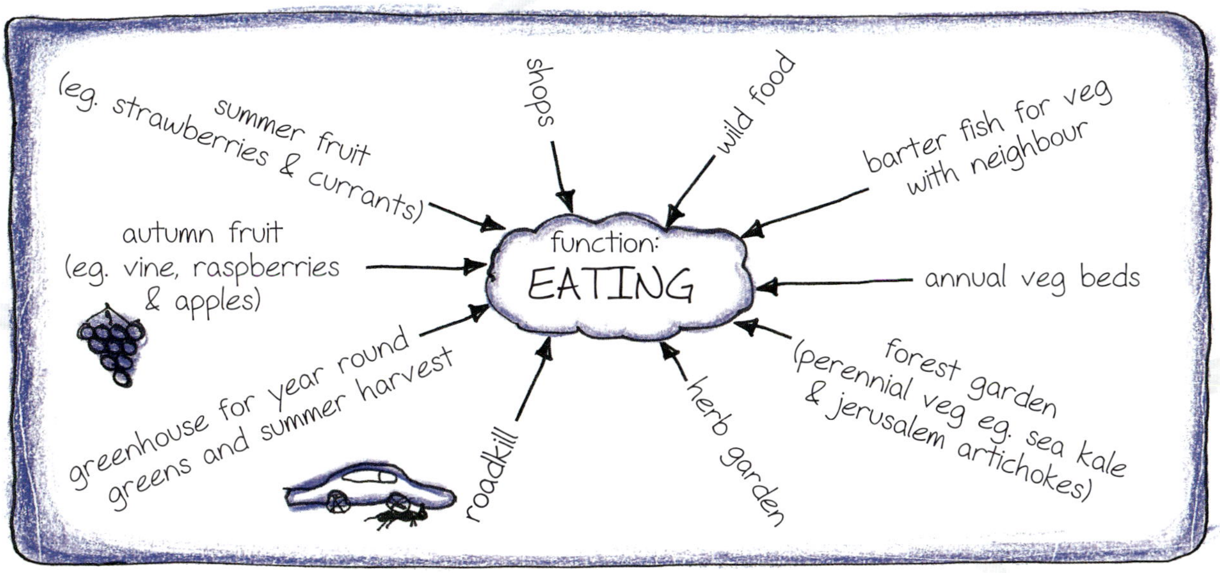

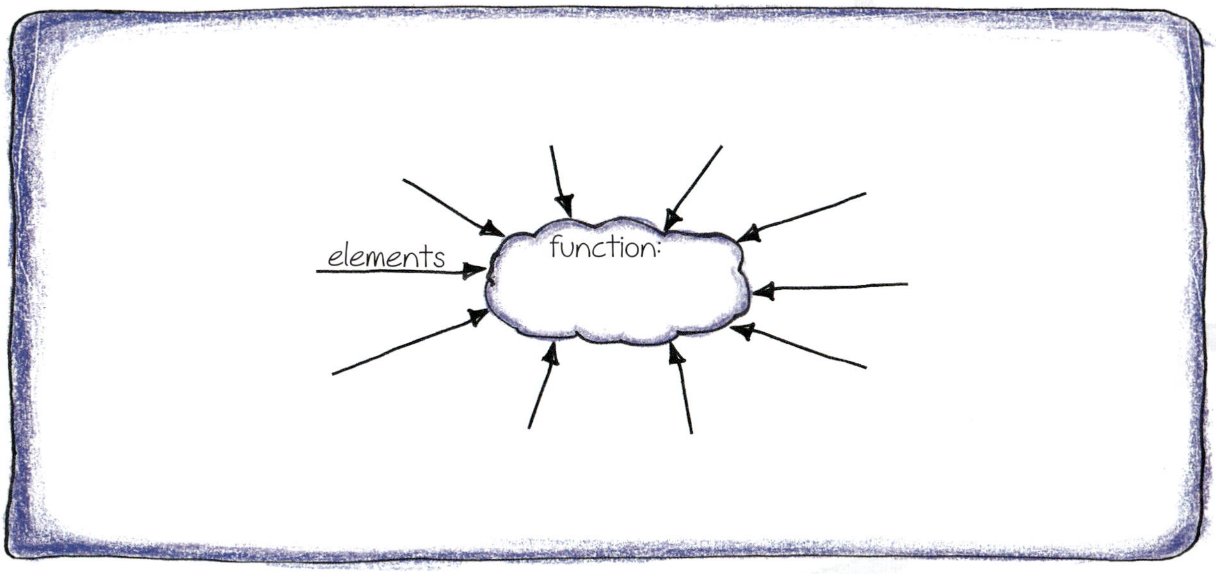

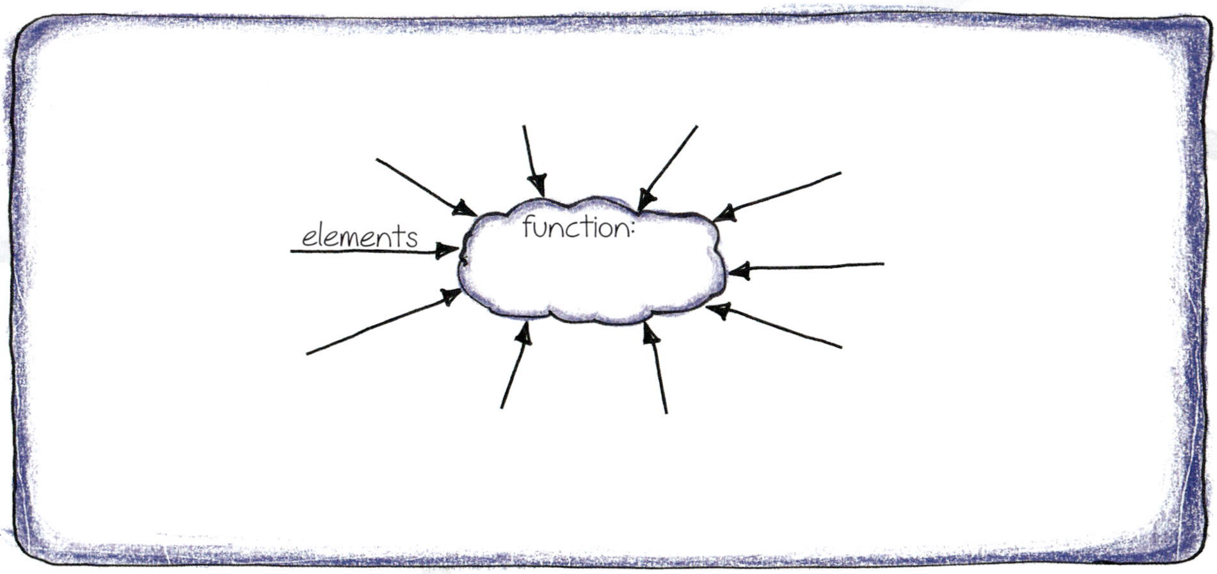

\+ POSITIVE

Which element of my design am I most pleased with for its multi function? Why?

Which function in my design do I feel is best supported by multiple elements? Why?

Evaluation

\- NEGATIVE

Which element is most challenging to work with? Why?

Which function is most challenging to work with? Why?

? INTERESTING

Unexpected or interesting outcomes from this exercise were?

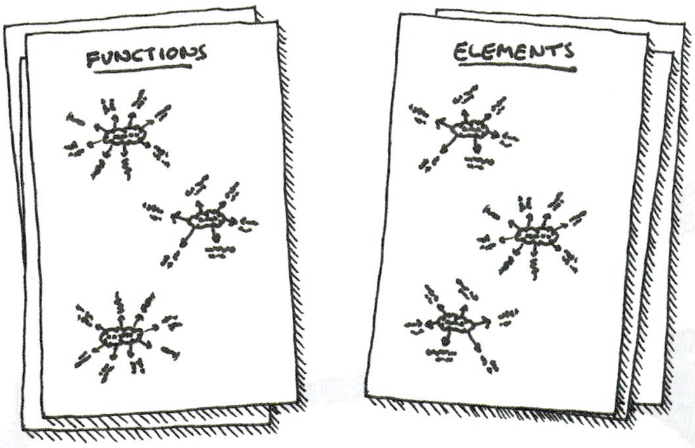

Synthesis

We are now going to combine all the landscape data and observations with all the analyses of elements and their functions into a coherent whole: a woven web. By now, you probably have a good sense of the ideal location and requirements for every aspect of your design. An excellent tool to achieve this synthesis between elements and landscape is McCarg's Exclusion Method. The basis of this design method is to ask where is LEAST appropriate for something to go, giving priority to protecting social and ecological values. This minimises intervention in ecosystems and places which are important to people's well-being. McCarg's book, 'Design with Nature,' is a wealth of knowledge from a hero of landscape sensitive design for those interested in digging deeper.

McCarg's method can be employed at an earlier stage in the design process. I've included it after the detailed analysis of all the elements, functions and their links deliberately. This is to minimise jumping to conclusions before understanding what the design elements will need and their impacts once in place.

Use your base map, sector analyses and transparent overlays (or pieces of paper to represent different constraints) and systematically check the locations for all your design ideas. Match the design elements to their best place by excluding areas for specific reasons. For example, when choosing a site to build our house, we ELIMINATED where it would most disrupt the ecology (deep soil); be too windy (to reduce inputs e.g. heating); too shaded (to maximise yields e.g. sun); and too visible (conserve neighbours viewscape).

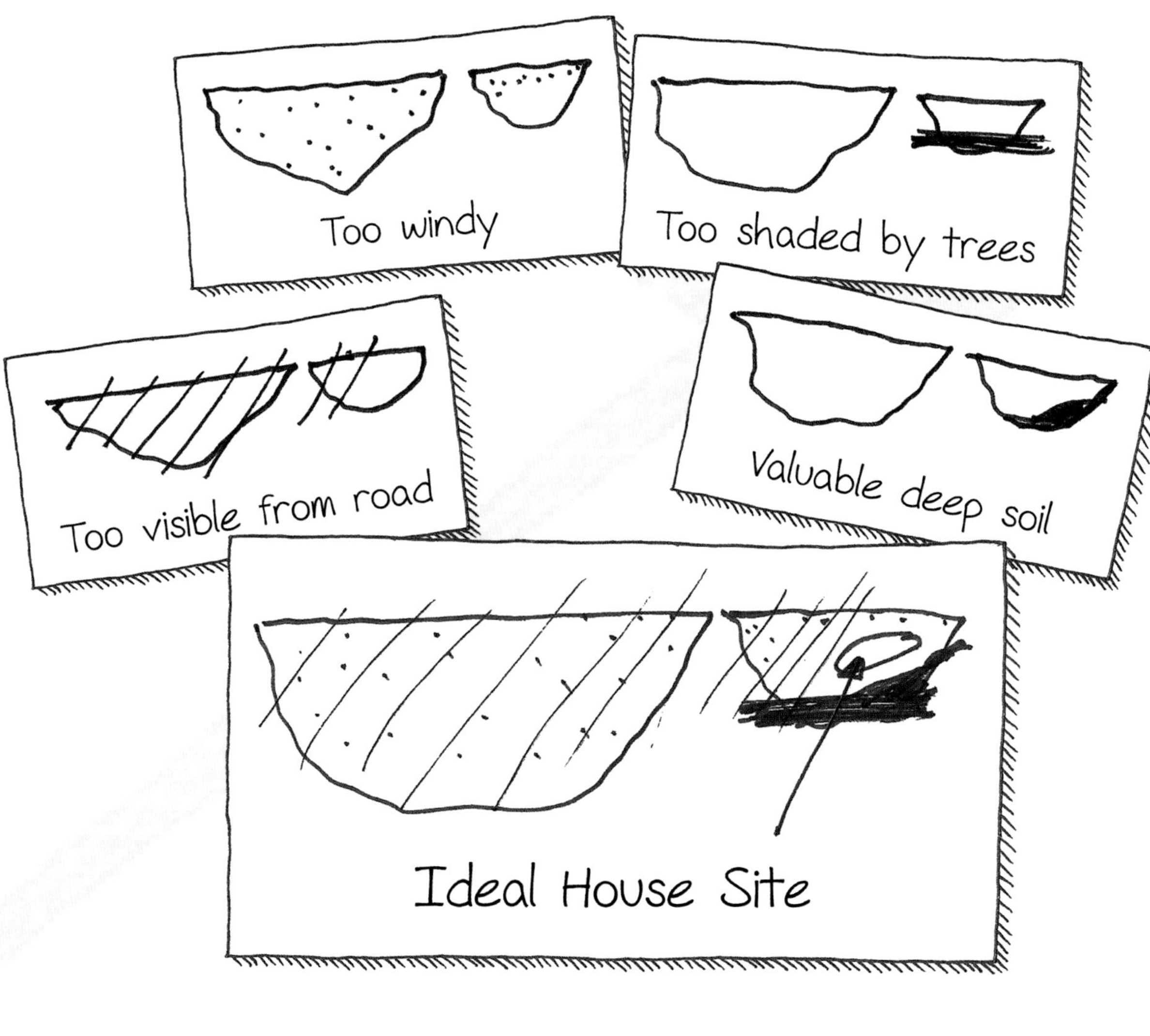

Here is another example for a simple vegetable garden:

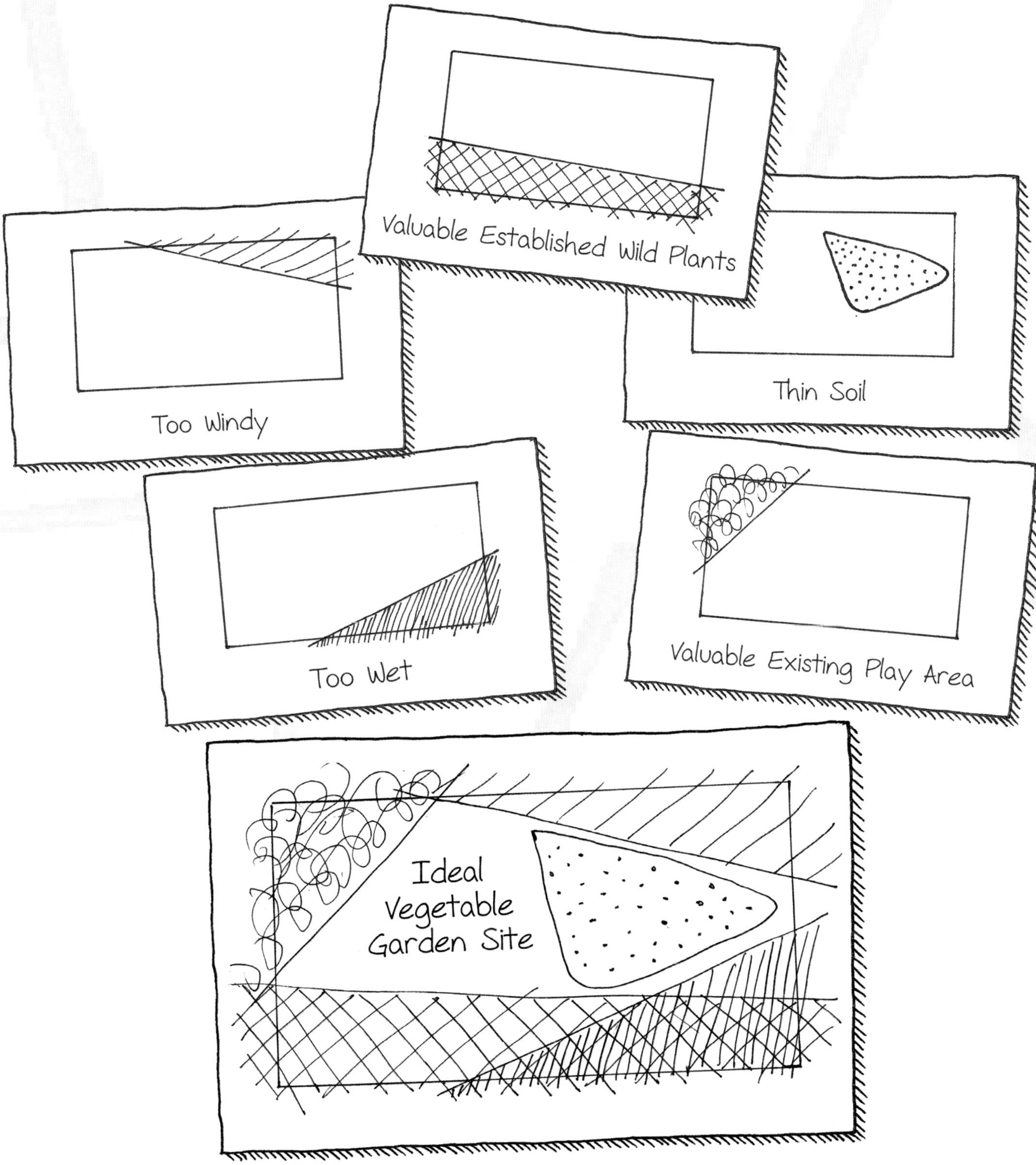

Remember, you can also take each element in turn and ask what the SWOC (strengths, weaknesses, opportunities, constraints) or PNI (positive, negative, interesting) potential is for it in any given spot. A quick comparison once the old supercomputer has checked through these processes usually yields the solution. The same applies if your conundrum is WHAT to include rather than where e.g. whether to have a shed or no shed; whether to have a big lawn or a small one; an orchard or a natural swimming pool...

PART 4:
Going Forth

The Design

Llongyfarchiadau! Congratulations, we are now on the home straight. The next pages introduce templates and guilds as a method to fine tune your design. This supports exactly where each element will go and how all their functions create a coherent whole.

Template Design Tool

We have seen how the elements can be placed in relationship to capture and store energy on many levels. A strong design tool to integrate this further is to make paper or card templates of every element in your design, to scale, from plan view (i.e. their footprint from the view above). This allows freedom to move them around and really look at the design from many angles, double checking all the connections. You can blue tack the templates onto your base map and allow time and space to check all that you have discovered is incorporated into your design.

You may have already been doing this or something similar on your base map and model.

When making templates of plants and trees it is best to do this at their MATURE size. In other words, an apple tree that grows to 5m diameter is represented by a template that reflects this rather than the 50cm diameter it is when you first plant it. This is because any landscape design, involving elements that change size, needs to see things as they will be, that way you anticipate future effects, such as shading out. In people designs, elements and functions change too over time, though we are less likely to use templates to represent them spatially.

Have a play with this bathroom design.

ELEMENT	INPUTS / NEEDS	OUTPUTS / YIELDS
Bath with shower	Water requires 1.5m x 1m of space plumbing not visible from windows	Water wet people and towels grey water and steam privacy and relaxation
Sink with cupboard underneath	Water 1m x 60cm plumbing easy access to cupboard	Water washed people/teeth etc. grey water and steam used bottles, toothpaste tubes
Seat with storage under lid	1m x 50cm A view to enjoy after the bath Place to sit, cut toenails etc	Used towels etc Relaxed people Toenails, empty lotion bottles
3 sub tropical plant in pots	60, 45 & 30cm diameter pots moisture/water heat/light air	Beauty, ambience help ventilate room leafy growth absorbs CO_2, exhales O_2
Laundry basket	50cm x 50cm Dirty clothes, fairly accessible	Dirty clothes, easily portable to washing machine etc.

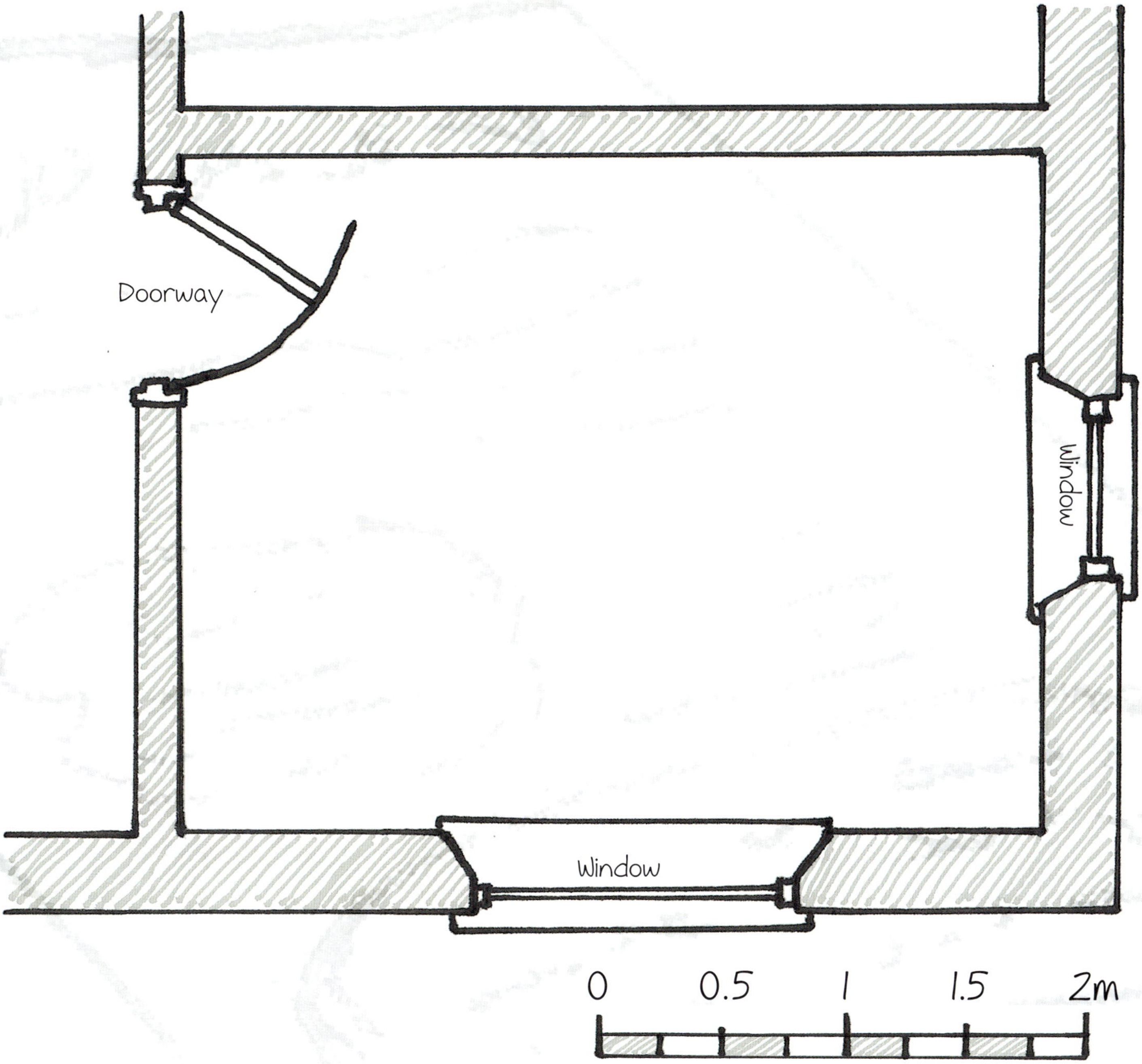

1:25 (4cm=1m)

Cut out the templates and move them around on the base map of the bathroom. On the opposite page are some salient inputs and outputs to consider in their placement. Note each element has multiple functions, except the laundry bin; maybe you can think of some way to diversify its use? Feel free to use the space beneath windows.

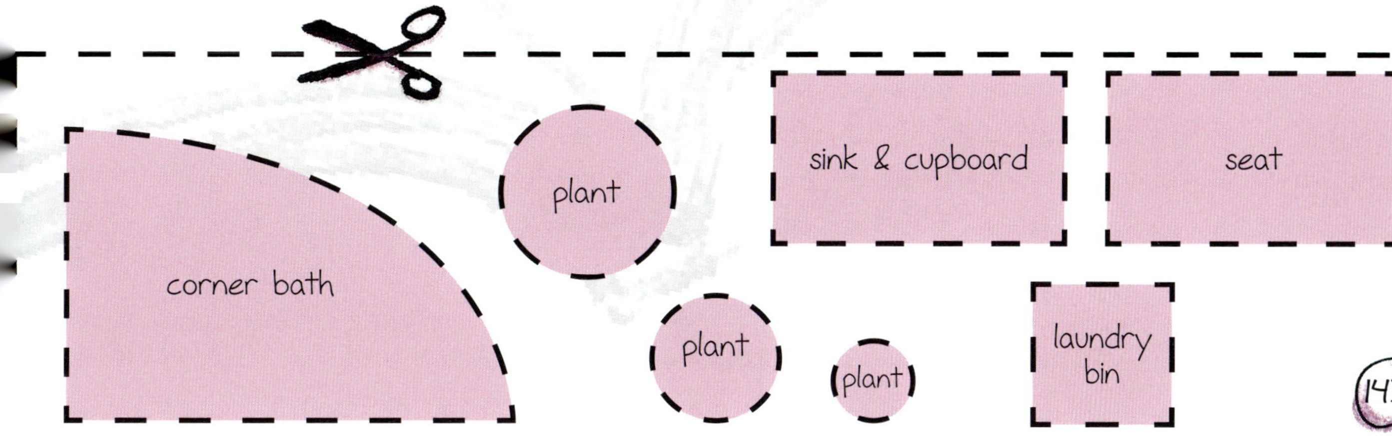

Creating Guilds

A guild is a group of elements that support each other through a network of beneficial relationships. It's the way the multiple functions of the elements take place in space and time effectively. Creating guilds within your design is a powerful tool to transform all our previous analyses into a coherent whole. By integrating rather than segregating your design elements, you use and value the diversity of all their characteristics and achieve the elegance and fluidity of energy and ecological efficiency we have been working towards. In a guild each element has a different role that mutually supports the others.

For example:

1. With edible crops, include other plants to attract beneficial insects and provide nutrients.
2. People that support the design physically and emotionally as well as the main designer and client.
3. Within a house this might mean spaces for relaxation and creativity, complementing work zones.

A guild also uses and values diversity in terms of elements occupying different roles in the space and light available and at different times of the year or throughout the day and over a lifetime.

For example:

1. A small patio design might have a vertical trellis, supporting autumn grapes and summer beans; a container with a small fruit tree and herbs in a window box.
2. People with skills to help set up a project and advise; someone who likes a bit of occasional heavy work and someone who enjoys small, delicate jobs when the design is underway, such as pruning and so on.
3. Areas for young children to play and an adjacent quiet space for adult socialising and working that allows for keeping an eye on them; an outdoor office that becomes a teenagers den in later years and so on.

By creating guilds we are designing from the patterns (of seasons, living and working flows) to details of how and where this takes place, modelling the integral relationships of how wild ecosystems and flourishing communities thrive. Using such a pattern language simplifies the complex, whilst retaining a high level of integrated interactions. This optimises the edge effect, ensuring marginal spaces, activities and even people are contributing to the overall potential.

To embed the idea of guilds, we'll do another template exercise with the example of diversifying a small orchard into a forest garden. Rather than segregating the tasks of pest control, fertility, irrigation, shelter and so on, we are going to arrange the elements to integrate those functions so that they meet all the needs of each element without having to bring in additional energy inputs to do so.

This template exercise is simplified for clarity, with minimal information on the base map and a brief summary of the inputs and outputs of each element. Bear in mind the sun and shade requirements for the lower growing layers in relation to the taller shrubs and trees. If you decide to use some elements as a wind break to create a suitable microclimate for the fruit trees, consider the optimal spacing of balancing shelter with shade. Martin Crawford's book, 'Creating a Forest Garden,' is a delight if you wish to follow this topic further.

Also, pay attention to the needs of access at harvest time, leaning through a spiky gooseberry to harvest apples is no fun. Crops from the root layer, such as jerusalem artichokes, require actual digging up, so leave plenty of space to avoid damaging the other elements, who by their perennial nature wish to stay rooted and undisturbed for many years to thrive.

It is fine to overlap the templates, as quite often it works well to site a smaller plant underneath a taller one; shade and sun requirements permitting. The trees can also be sited with their canopy going beyond the outline of the map. Finally, the mint, wild medicinal plants and strawberry templates in the exercise are represented as 'drifts', i.e. where an initial planting is designed to spread.

Feel free to add more features, e.g. access or different plants.

FOREST GARDENS

The concept of forest gardening was pioneered by the late Robert Hart. He observed the efficiency and productivity of tropical food gardens, compared to monocultural crops in rows, and transplanted the idea to Britain.

The basic wisdom is to emulate the climate's most thriving and stable ecosystem; in our case a forest or woodland. In cool temperate climates, we have much less intense sun than in the tropics; our diversity and yields are actually taking place on the woodland edge or where glades exist within a forest.

The elements in a forest garden are selected for human needs; plants which yield edible, medicinal and other useful outputs.

The main characteristics or principles to include are:

- Model the layers of a woodland: canopy/large trees; small trees; shrubs; climbers; herbaceous; groundcover; roots and fungi.
- Allow more space and light compared to a normal orchard for the lower layers to thrive.
- Select perennial and self seeding varieties (more resilient, less work, undug /wild soils).
- Design elements in guilds (fostering the benefits mentioned above e.g shelter, fertility and pollination).

LAYER	ELEMENT	INPUTS / NEEDS	OUTPUTS / YIELDS
Large tree	PEAR	Shelter Pollination Fertility Pest control	Fruit Shade Microclimate effects
Small tree	ELDER	Happy in shade Access for harvesting Space	Fruit and Cuttings Shelter Shade Habitat
	AUTUMN OLIVE x3	Happy in shade Access for harvesting Space	Fruit Habitat Fixes nitrogen for other plants Good windbreak
Shrub	BLACKCURRANT x3	Happy in sun or partial shade Access for harvesting Some fertilising helps fruit	Fruit Habitat Good for cuttings for new shrubs
	GOOSEBERRY	Happy in shade Access for harvesting Some fertilising helps fruit	Fruit Habitat Has spikes
Climber/ Vertical	OUTDOOR GRAPE	Tree or structure to climb Light Air Some shelter Pollination Regular pruning	Fruit Lots of leaf litter Shade Habitat
Herbaceous	COMFREY x7	Sun Room to spread Harvesting	Nitrogen, potassium and phosphorus Medicinal Bee fodder
	WILD EDIBLE AND MEDICINAL PLANTS	Some sun Some maintenance of species balance Space	Minerals Attracts beneficial insects, Beauty Habitat Edible and medicinal
Ground cover	MARJORAM	Sun Space Air	Attracts insects Protects soil Edible
	MINT PATH	Happy in shade Access for harvesting	Repels pests Edible Protects soil
	STRAWBERRIES	Sun Harvesting Mulching No trampling Pollinating	Fruit Habitat Protection from erosion, Water retention
Root	JERUSALEM ARTICHOKES	Happy in shade Access for harvesting Space Grows tall	Gives some wind shelter, Edible crops Habitat
Fungi	SHITAKE MUSHROOM LOG	Damp and shade Deadwood Initial innoculation	Habitat Edible crops Mycorrhizal function shares nutrient and water

Cut these out...

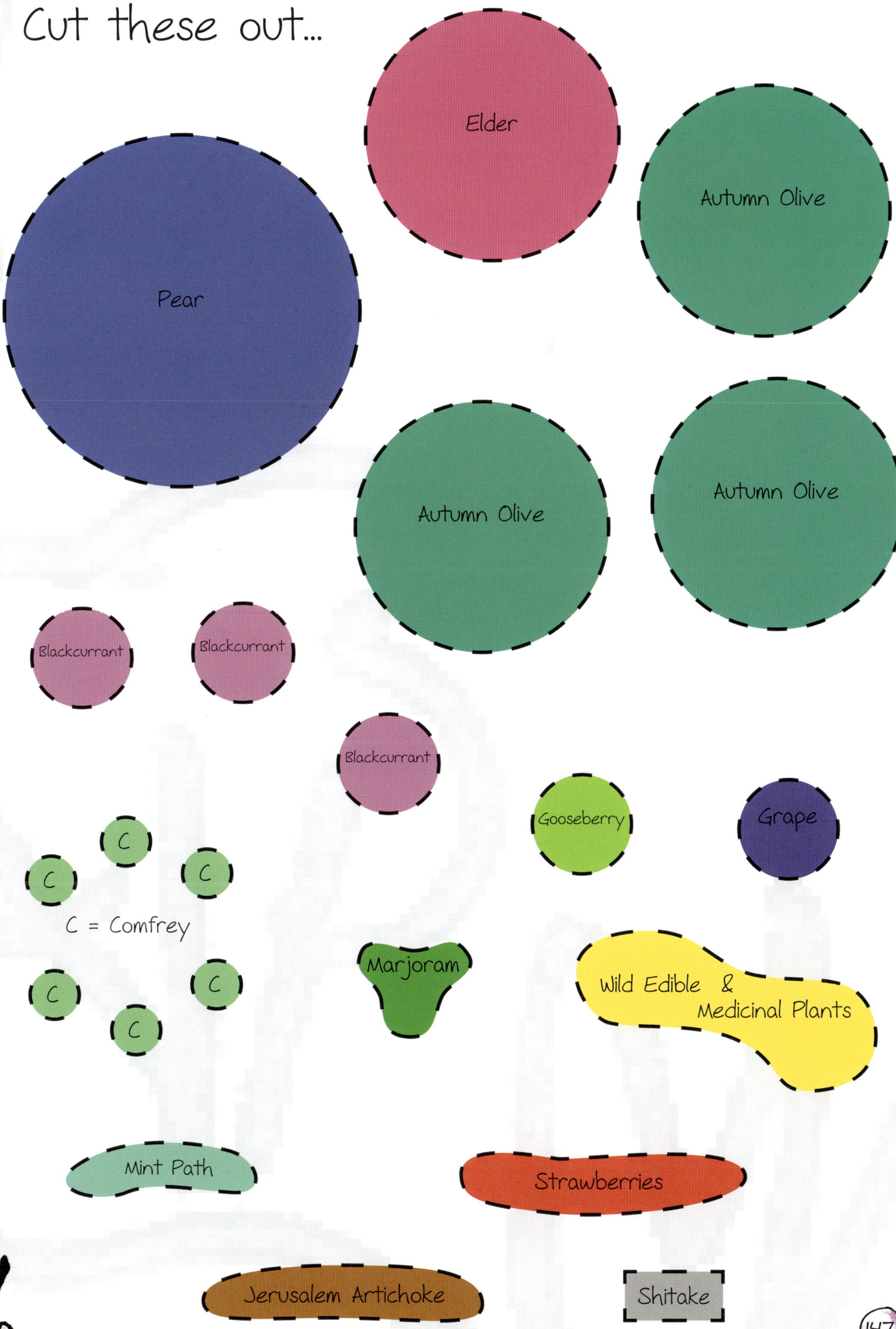

Elder
Autumn Olive
Pear
Autumn Olive
Autumn Olive
Blackcurrant
Blackcurrant
Blackcurrant
Grape
Gooseberry
C
C
C
C
C
C
Shitake

BASE MAP:
Orchard being diversified to a Forest Garden with Guilds

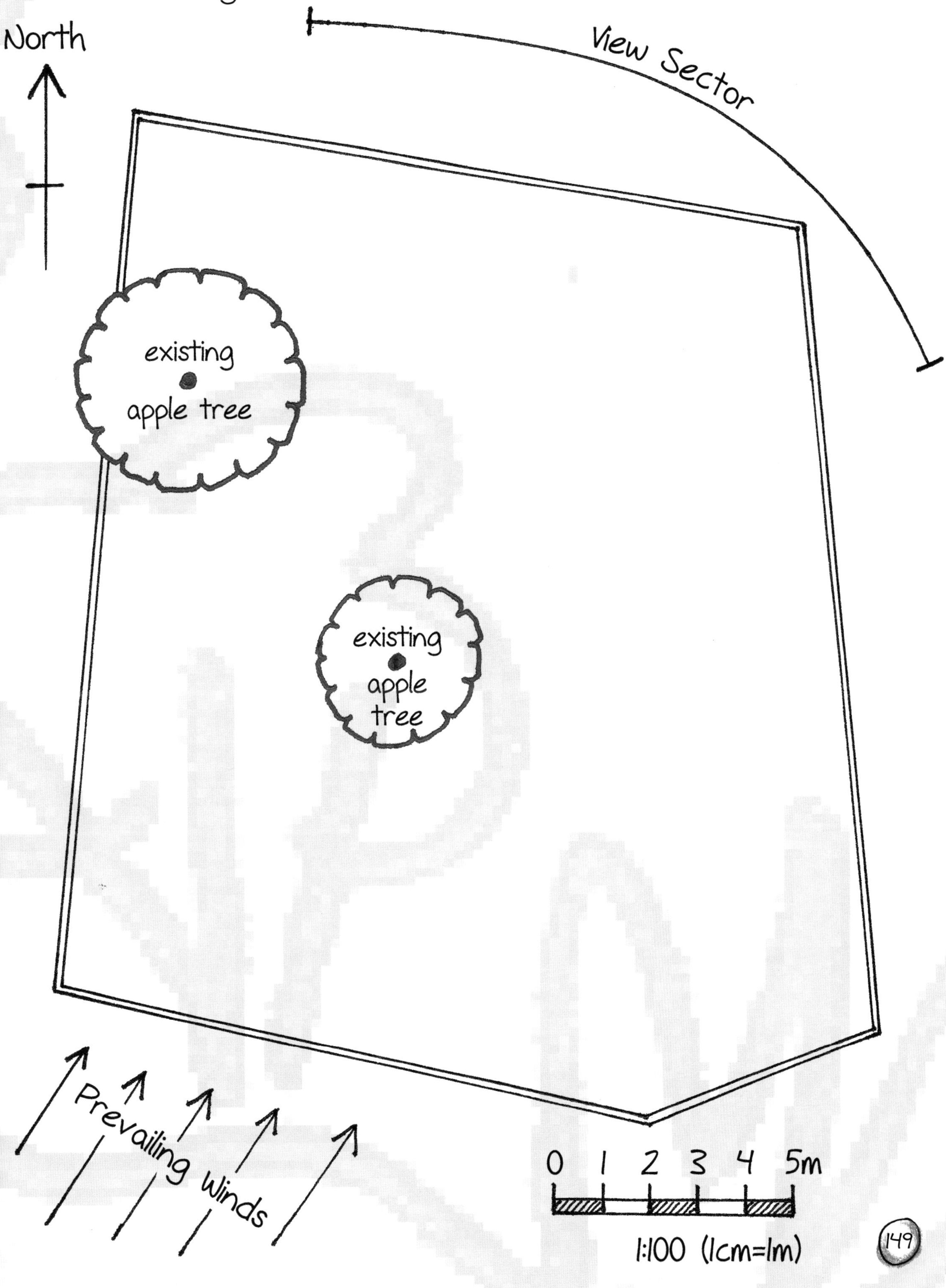

When you've finished this exercise, apply the guild tool to as many aspects of your own design as you can dream up. Give plenty of time to this, experimenting with different ways of assembling the elements. Work with one zone or area at a time. If your design is for a small place, it may well be the whole design is one guild.

Remember to draw the templates of each element to scale and use them in conjunction with the base map. If you are working on a model, a 3D object works well and is fun. We found a standard 'go-go' or 1cm square lego piece was a good scale to represent a fruit tree on our model.

Use the questions below to keep you on track for the physical elements of your design, such as plants, trees, buildings, gazebos and so on. Fill in the box when you are satisfied with your guilds.

1. I applied the guild tool to

2. How did the different elements diversify the design over time?
e.g. the seasons, the day, years...

3. Which spatial niches and edges does your guild occupy?
e.g. vertical, underground, shrub layer...

4. Name some of the beneficial interactions your guild has created...
e.g. providing fertility, attracting pollinators, microclimates for shelter and sun.

Now select a people aspect of your design to apply the guild pattern to. This ensures resilience and efficient implementation of your vision and the long term maintenance of energy, skills and enthusiasm required for this. For a people centred design, written notes or a drawing will work to allocate roles in a guild. Find your own method.

1. WHO WILL MAKE THIS DESIGN WORK OVER TIME

(e.g. builders, advice from plant nurseries; Uncle Henry and his £50 for Gran's sun seat; yourself for pruning back the rambling rose once it's established...)

2. WHAT DIFFERENT ROLES WILL BE IN PLAY THROUGHOUT THE DESIGN'S LIFE

(e.g. granchildren visitng (sandpit); Dad to mow lawn; Gran to enjoy it; space for a veg garden if I move in as a carer...)

3. NAME SOME OF THE BENEFICIAL INTERACTIONS BETWEEN THE PEOPLE OF THIS GUILD?

(e.g. low maintenance child care allowing adults to relax and make visiting Gran more likely; something active for me to do nearby...)

Applying Permaculture Principles

Before drawing up the final design, I like to do a final check by going back through the permaculture principles to make sure I have covered everything as best I can. My view is that great minds observed these principles as universal to thriving systems, themselves made by the ultimate designer, Nature. If I know I have met these principles to the best of my ability, this gives me great confidence to proceed with my design. Whilst I do this, I am also having a final check in with the overall vision of the design and its suitability for all those concerned before I make an implementation and maintenance plan.

Make a note of how your design applies all the permaculture principles, however you interpret them. I have included examples from my own experience, in case you are stuck for ideas, in the Additional Resources at the back of the book. These examples show how the principles were met on a smallholding scale, however they can be applied on any scale, even if you have no control of your outside surroundings. For example, observation of wind leading to catching and storing energy with a windbreak:

~ hundreds of trees can be a windbreak, sheltering homes and farm crops

~ small shrubs or a trellis on a patio or door step in a small garden or flat

~ thicker curtains where cold draughts are observed!

OBSERVE AND INTERACT

"Beauty is in the eye of the beholder"

~

Continuous observation of nature (including ourselves) is the foundation of good design and fosters minimal interference with natural cycles.

Name the ways in which observation of your place, the people involved and yourself, have informed your choices.

CATCH AND STORE ENERGY

"Make hay while the sun shines"

Energy flows, such as water, wind, nutrients and sun, constantly move through the landscape. To minimise harvest of fossil fuels and recognition of their finite nature, we can reinvest and preserve energy in many ways. Finding ways to conserve our own energy is vital too.

List how your design does that here.

OBTAIN A YIELD

"You can't work on an empty stomach"

This principle reminds us to design a system for self reliance at all levels, not just for food crops. A good yield promotes the system that generated the yield and creates positive feedback loops.

Name the ways in which you expect a yield in the spheres of food, ecology, energy, cycling of waste, socially, emotionally, educational and any others you can think of.

APPLY SELF REGULATION AND ACCEPT FEEDBACK

"The sins of the fathers are visited on the children unto the seventh generation"

~

With understanding of how negative and positive feedback works in nature and how their effects are often delayed over time, we can design systems that are more self regulating, reducing work and intrusive management.

In what ways have you learnt from your design so far, and how do you plan to in the future? Name any key insights from research that led to an adapted design too.

USE AND VALUE RENEWABLE RESOURCES AND SERVICES

"Let nature take its course"

~

Renewables are resources (e.g. from plants, sun, wind and water) we use without depleting them. This also includes activities and interactions in the human sphere that regenerate us. By following this principle, we lessen the demands of consumption on the planet and often means matching inputs to outputs.

How has your design harvested and supported renewable resources?

PRODUCE NO WASTE

"A stitch in time saves nine"

As we saw in the Input-Output analysis earlier, thriftiness and creative thinking can help us tread more softly on the earth and create abundance, instead of wasting resources and opportunities.

Think of all the ways your design saves time, energy, materials and avoids waste and pollution. List them here.

DESIGN FROM PATTERNS TO DETAILS

"Can't see the wood for the trees"

This principle rests on the idea that complex systems begin as simple ones and a focus on details prematurely tends to make it harder to understand the whole. Patterns allow us to make sense of a myriad and evolving world.

Name the patterns you have incorporated into your design.

INTEGRATE RATHER THAN SEGREGATE

"Many hands make light work"

The connections between elements is integral to the functions they perform, whether ecological, social or personal. Designing in beneficial relationships models ecosystems and creates resilience and efficiency.

Make a note of all the ways you have integrated elements and their functions by their placement in the design and in making sure they are are met in multiple ways.

USE SMALL AND SLOW SOLUTIONS

"The bigger they are, the harder they fall. Slow and steady wins the race"

Slowing down and increasing self reliance makes us more observant and in harmony with our surrounds. This principle emphasises that excessive growth leaves us vulnerable and invites us to do our projects on a human scale, allowing time and energy to observe, learn and feedback the results of our endeavours before making more decisions.

What small and slow solutions does your design encompass?

USE AND VALUE DIVERSITY

"Don't put all your eggs in one basket"

The greater diversity of forms, functions and interactions you can build into your design the better. Diversity is the basis of thriving natural systems and provides a dynamic and interesting expression of life that is resilient compared to the vulnerability of monoculture.

Name the ways you diversified your design ecologically (choice of plants and different habitats); socially (inclusion of various people and groups); and personally (in your approach to this design process, the practical application of it and in other areas of your life that will become involved).

USE EDGES AND VALUE THE MARGINAL

"Don't think you are on the right track just because it's a well trodden path"

Where two systems meet, a greater diversity of life forms flourish. Coasts, wetlands and pond shallows are lively edges. Plants find sanctuary in cracks in concrete. Rich cultures spring up in diverse, ethnic centres and creative solutions appear in marginal spaces following economic difficulties.

Name the edges your design inhabits, across time and in the ways you plan to use the space. Include any social or personal edges you have incorporated too.

CREATIVELY USE AND RESPOND TO CHANGE

"Vision is not seeing things as they are but as they will be"

This principle recognises long term sustainability requires flexibility and adaption by ecosystems, in our own lives and of whole cultures. It takes us full cycle back to the principle of observation.

Consider global and national issues; the potential change in your region and local area. Consider the people involved in your design and possible change in personal circumstances. In what ways do you think your design responds to potential change in a constructive way?

Evaluation

+ POSITIVE

What works well for you when applying the permaculture principles as a design tool?

- NEGATIVE

What's challenging when applying the permaculture principles as a design tool?

? INTERESTING

Did this exercise yield any unexpected or interesting insights?

Concept Design Map

Making a Concept Map, to scale, is a useful way to see at a glance the design WITHOUT squashing in all the details. Draw the outline of your site below. Now, represent the various distinct areas you plan to create (e.g. with broad shading or zones) and the main elements of your design to create a concept map. Depict any key information, visually and simply, that is central to your design.

Detail Design Map

The Detail Map of the whole site benefits from being drawn onto a large piece of paper, making it easy to understand and follow. Specific areas that have lots of elements (e.g. a forest garden) or diverse functions (e.g. an outdoor classroom) or that require detailed measurements (e.g. a greenhouse) benefit from having their own detail map. Photocopy or draw an aspect of your whole design below as a detail map.

Implementation

Yes! It is almost time to get stuck in and make our designs physically manifest. Our initial vision is now a detailed map, like a seed ready to burst into life. In our journey through the SADIMET design process, the Implementation and Maintenance steps are here to remind us to pay attention to turning our design into achievable goals. Many great ideas falter through inadequate preparation or insufficient planning of how to follow the dream through in practice.

The Rudyard Kipling wisdom, in the quote below, is a good way to structure a strategy for implementing your design – proper planning at this stage frees us up again to get busy, be active in the actual 'doing' of the design and have fun and satisfaction rather than indecision and overwhelm.

> *"I keep six honest serving men*
> *(they taught me all I knew);*
> *their names are What and Why and When*
> *and How and Where and Who."*
> Rudyard Kipling

The table opposite is a method to make a chronological (i.e. first things first) plan of how you will implement your design. I tend to make a rough plan and timeline, then transfer the information onto the table. You may well wish to make more detailed 'To Do' lists for the different tasks.

Remember, you can write each task and time phase onto card first, akin to using templates, moving them around on the table until you are happy with the sequence. You can apply the analysis tools of SWOC and PNI to check which order, or whom, or how and so on suits your implementation success the best.

Er... you can also group your tasks into guilds for complimentary beneficial relationships and draw on the permaculture principles to check your approach to implementing meets the criteria of a sustainable system...

When you have done the above and are happy with your structure, fill in the table opposite for a clear pathway of how you are going to proceed. For a group project, copy the table onto a large piece of paper and display somewhere everyone can see it easily.

WHAT	WHY	WHEN	HOW	WHERE	WHO	COMMENTS
task, activity or element to be implemented	reason for position in the timetable	timescale of implementation (exact or estimate)	method of achieving this; process, people, money	specify location within the design	who is doing what	things to remember/ flag up

Maintenance: Troubleshooting

As we know by now, the map is not the territory. The reality of maintaining a plan of action frequently throws up new challenges. This exercise is to support and maintain the necessary energy to fulfil your design by troubleshooting for potential problems. A core idea of permaculture is to transform problems into solutions.

This attitude can transform an overwhelming challenge into a rewarding opportunity. When we got our land, it was a marginal, unwanted, steep field. The problem of making it productive was very daunting. I examined my initial reasons for resonating with permaculture, and remembered its power to repair denuded landscapes. The problem became the perfect opportunity for me to do just that - permaculture is exactly the solution to transform such a place into a fertile haven. This approach addressed my fears and motivated me.

	POTENTIAL / ACTUAL PROBLEM
Materials / resource shortages	
Money flows	
Skills shortage	
Personal energy	
Resistance from authorities (landlord; your own conscience; governmental; parents)	
Family commitments	
Work commitments	
Ideas on paper impractical on the ground	
Imbalance of fun to hard slog	
A new passion or project comes up	
Struggle to maintain areas established previously	
Unexpected problems (e.g. neighbours, illness)	
Other	

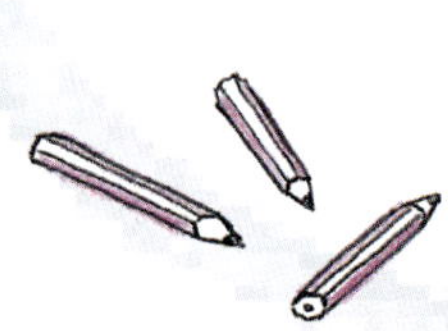

To maintain your design, look at what might come up to interfere with implementing your goals. Draw on your own and others experience. I like to name 3 solutions per 'problem' to ensure a strategy will be found.

If your design is quite simple, with relatively few implementation phases or tasks, you can troubleshoot the whole design on the table below. For a project that has distinct areas, with different characteristics and needs, being implemented at different times, I recommend making copies of the table and completing the exercise below separately for EACH area or task. For example: a garden project that aims to build a patio and outdoor play area; followed by establishing a vegetable garden, would benefit from each being addressed individually.

Keep track of the Maintenance stage of a design whilst it is in progress; refer back regularly to how you planned to Implement your ideas. If some aspects, (or all), aren't working, be gentle on yourself and deal with it further when you evaluate and tweak your plans.

SOLUTIONS	RESOURCES REQUIRED	DO DIFFERENTLY NEXT TIME

Action Learning Plan

A final resource I'd like to offer is an Action Learning Plan (ALP): a plan of action to support being an effective designer. We always do an ALP when designing or at the end of a course. The notion of a continual cycle of learning – action – observing effects – learning and so on, is integral to the systems approach of permaculture. Making such a plan is an opportunity to expand ourselves and those around us and to reach the potential of the design.

Your A.L.P may be

~ Focused directly on skills or research required to carry out your design

~ Related more indirectly e.g. topics that came up, though not essential for the project

~. Entirely new territory as this design process ends and is no longer your main focus.

Here's my current ALP for making a Zone Zero art cabin in the garden.

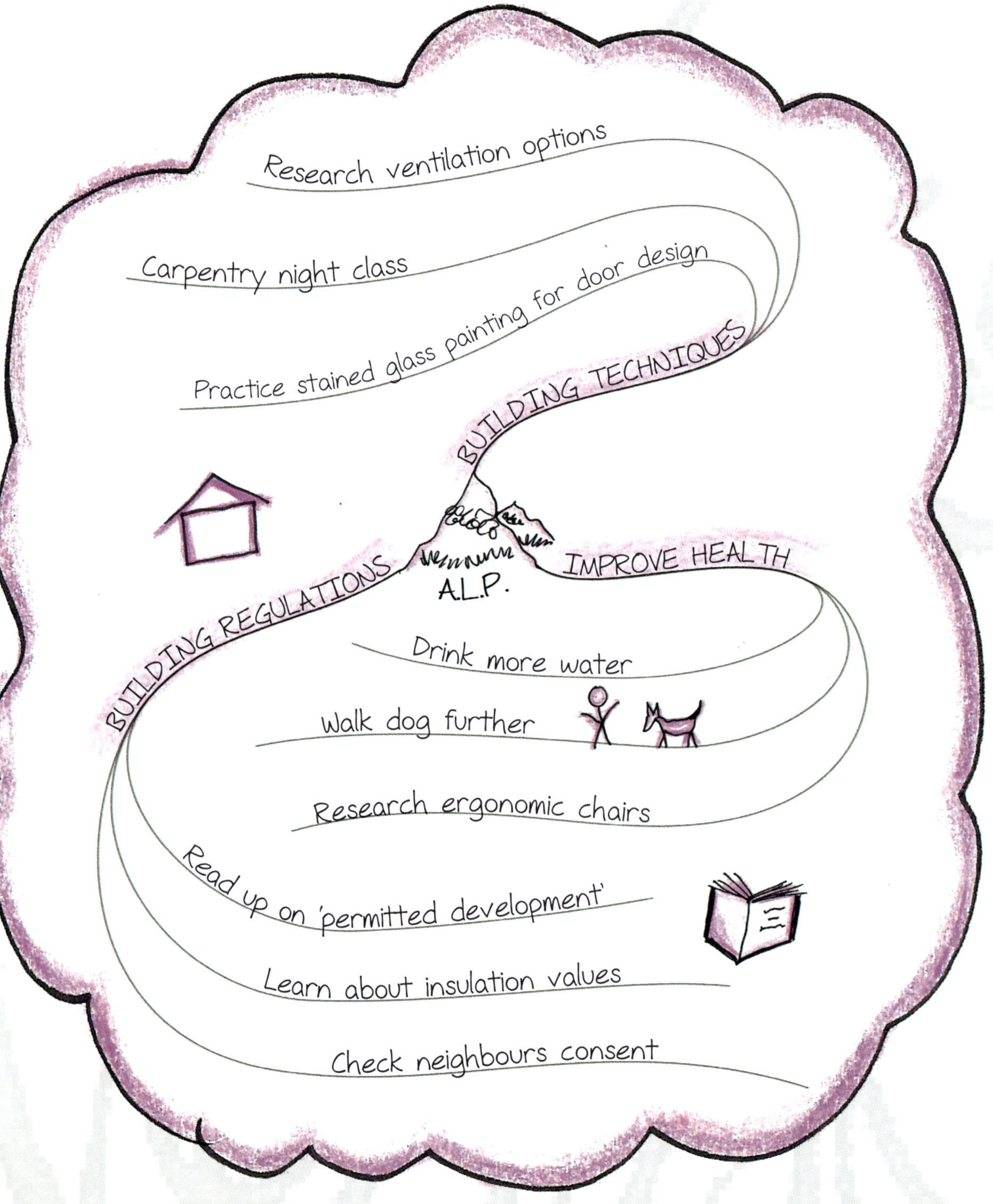

Make an ALP with the mindmap tool, developed by Edward De Bono, which models the pattern of how our brains think, using colour and symbols to embed the info.

Evaluate and Tweak

This a simple method of keeping our experience active and useful; cycling feedback into future activity and learning from mistakes. Choose a suitable timescale to factor in a regular evaluation and tweak. A quick appraisal at the end of a day to prepare for tomorrow's tasks; a weekly review; at the start or end of a season (your fixed point photography records are good to dig out here) and so on.

Today's Date	Phase within Overall Timescale	Specific Tasks Evaluated	Relevant Design Element(s) / Zone	What's Going Well?

~ Enjoy the table below to keep track of your evaluations and tweaks.

~ Make copies to keep the process going.

~ Think of your own and other people's satisfaction; energy; time; and budget.

~ Include anything else relevant that comes up.

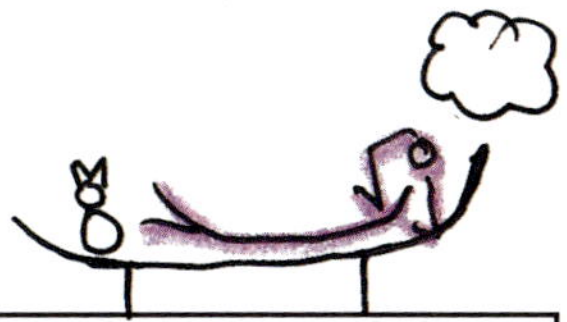

What's Challenging?	Check with Long Term Vision: Still Relevant?	What Can I Do Differently or Adjust?	Resources Required

An Ethical Foundation

Way back in Part I we found the core of permaculture was Observation, Principles and Ethics and the Design Process. As a former political economy student, I had been trained to be neutral and put aside my own ethics, yet deep down yearned for the meaning they give. I was pleasantly surprised to find permaculture has an ethical basis, a heart in amongst all that intense hard science, complemented further by the call to practical action. Actual, real, achievable solutions that are people and ecologically friendly.

The permaculture ethics are:

EARTH CARE:
of natural systems (e.g. soil, water, air, biodiversity, all species, the earth itself as a self-regulating whole).

PEOPLE CARE
with a focus on societal patterns (e.g. families, communities, nations and future generations).

FAIR SHARES
recognises limits to consumption and is a directive to redistribute surplus (in any form) wherever possible.

An excellent resource focused on bringing these ethics to life is Looby Macnamara's book, 'People and Permaculture.' It explores how to restore personal, social and planetary well-being and is a celebration of the potential of the human spirit.

I understand the permaculture ethics through the lens of Schumacher's concept of Soil, Soul and Society. Soil pertaining to earth care; society to people care and soul to fair shares. In our contemporary world, the interdependence of goods and services spread across the globe makes it extremely complex and challenging to live a life with no harmful impacts. It is impossible to be perfect and not make compromises to earth and/or people care unless we somehow reject any manufactured good, travel by foot, only eat food that is wild or we have grown and somehow not incur governmental expenditure on our behalf. A tall order and undesirable for most people.

Schumachers' principle of Soul resolves this dilemma through flagging up our response-ability as sovereign individual souls to employ discernment. This gives us a route as to how to observe the permaculture ethic of Fair Shares in a modern context. Here, discernment means being mindful of our lifestyles, consequent impacts and resource use and then being able to live with that consciously. This way we can focus on doing good and creating beauty and not be overwhelmed or feel despair.

Spend some time casting your mind over your whole design; the process, the result and the future actuality of it. Write in each shape how your intent meets the permaculture ethics. Note, the shapes overlap and influence each other as befits an integrated systems approach.

Here's an example of how I employed the permaculture ethics in designing this book

EARTH CARE

- Whole book devoted to supporting earth care
- Minimise fossil fuel use and pollution, through eco friendly printing
- Maximise activities for readers to appreciate nature first hand.

PEOPLE CARE

- Pay attention to exercises being accessible and fun
- Balance between my work and family life whilst making the book
- Credit other people's work wherever possible.

FAIR SHARES

- On balance, I feel the ecological benefits this book will foster outweighs the resource use and consequent pollution to create it.
- Allow people to share the work for free
- Continue to be as 'eco' as possible in my own life.

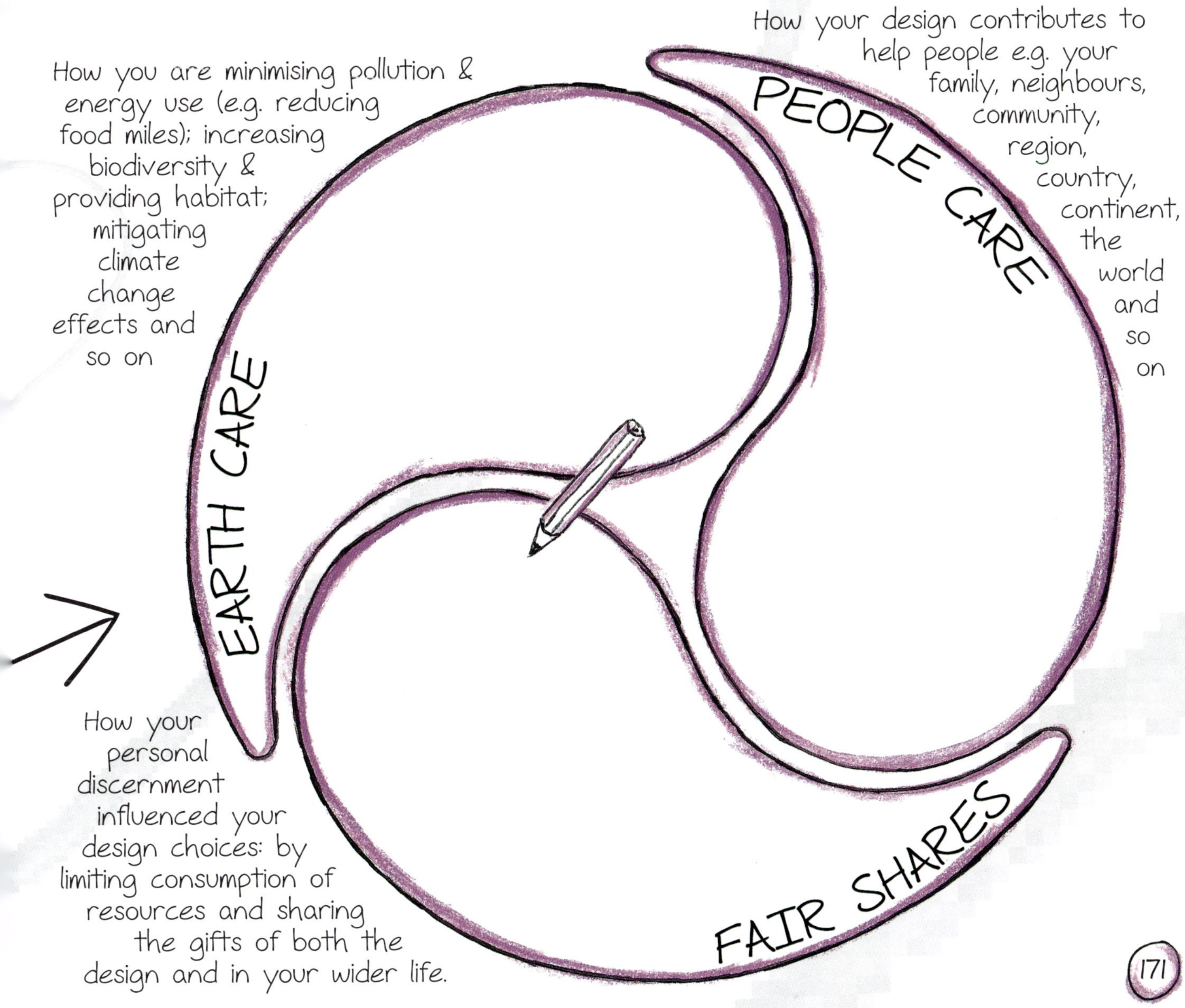

Next Achievable Steps

Our journey has come to an end. It has been my profound pleasure to be your companion through this process. All that remains, is for you to pledge your next steps to bring your design to life.

My next ACHIEVABLE step to action my design in the coming...

24 hours:

Week:

Month:

This book is built on the experience of ten thousand interactions with visitors, students and other facilitators over the span of ten years of teaching. Thank you one and all, especially to the land, for giving me the opportunity to become a neo-indigeneous Being in this day and age.

To keep this positive learning spiral ascending, please complete the PNI evaluation of the Permaculture Design Companion and email your thoughts to

jasmine@beingsomewhere.net

Whole Process Overview of working with this Permaculture Design Companion:

+ POSITIVE

- NEGATIVE

? INTERESTING

Good Luck, Go Well and Go Forth

Additional Resources

Landscape Survey

EXISTING BIODIVERSITY	CLIMATE
ecosystems (eg. hedgerow, marsh etc.) plant species animal species (include pests) old trees what's growing... ...how well	rainfall (mm/year) average temperatures first & last frosts prevailing wind patterns in different places
LANDFORM	**SOIL**
elevation (m above sea level) slopes (erodibility / condition) aspect (which way slope faces)	pH/ nutrient tests indicator plants profiles texture variations
MICROCLIMATES	**WATER**
frost pockets sun traps shady places windy places sheltered places damp / wet places well drained / dry places combined effects	SOURCES: mains existing taps roofs and downpipes greywater springs streams ponds possible pond / water storage sites NEEDS: irrigation (crops/garden) animals aquaculture domestic (drinking, cooking, washing) commercial e.g. food processing

<table>
<tr><th>SERVICES</th><th>BOUNDARIES & INTERACTIONS</th></tr>
<tr><td>underground
overhead cables
sewage systems
land drains
mains electric
off-grid electric</td><td>north
east
south
west</td></tr>
<tr><th>STRUCTURES</th><th>BEAUTY</th></tr>
<tr><td><u>BUILDINGS</u>
condition
energy potential (solar, rainwater etc)
relationship to land (current purpose if any)

<u>WALLS & FENCES</u>
aspect
condition</td><td>views from site
within site
off site from outside</td></tr>
<tr><th>ACCESS</th><th>MISCELLANEOUS</th></tr>
<tr><td><u>PATHS, ROADS, TRACKS, GATES, BRIDGES, DOORS ETC</u>
siting
type and frequency of traffic
steepness/ erosion potential
condition
seasonality
desire lines (where folk naturally walk)</td><td>resources
hazards
special areas (archaeological / sacred)
off site services (eg transport/ markets)
unique features</td></tr>
</table>

Managing Complexity

POSSIBLE FACTORS TO CONSIDER IN A DESIGN

Land

taps

water sources

sun

rivers

views

shade

frost

ponds

springs

land actvities on adjacent properties

wildlife corridors

pipes

energy

power lines

contours

servicies

soil pH condition

biodiversity

boundaries

suntraps

animals

unusual features

invasive species

cess pit

access

slope

wind

walls

plants

big trees

bedrock

height above sea level

micro climates

materials

dominant features

People/Society

regulations e.g. planning sanitation legal requirements, structural engineers etc

neighbours

local areas

tree protection orders SSSIs AONB

market research

bylaws e.g. fishing rights public footpaths rights of way

family

other relevant projects

ruling bodies/ members/committees

soil association

land registry

Personal

motivation

ideas

personal qualities

background

finances

health

contacts

commitments to family and work

self-esteem

friends

areas of life in transition

ancestry

plans

time

emotional well being

skills required/ acquired

dreams

passions

creative

Brief Comparison of Design Frameworks

DESIGN FRAMEWORK	WHAT WORKS WELL	WHAT IS CHALLENGING
<u>SADIMET</u> Survey Analyse Design Implement Maintain Evaluate Tweak	Thorough, gives confidence for complex designs with existing biodiversity. Using full landscape survey metaphorically on people designs.	Permaculture jargon can put some people off. More depth than necessary for very simple design situations.
<u>OBREDIMET</u> Observation Boundaries Resources Evaluate Design Implement Maintain Evaluate Tweak	Emphasis on boundaries useful when there are interactions between distinct zones in the design. Suits a less intensive survey of the landscape (e.g. little biodiversity or as a general assessment.)	Requires lateral thinking to apply to the people aspect of design. To remember that observation runs throughout a process and not merely the first step.
<u>CEAP</u> Collect Evaluate Apply permaculture principles Plan	Useful when a full landscape survey not necessary. 'Collect' is a versatile tool to catch and store useful info. Applying the PC principles covers all angles.	No reminder to evaluate and feedback learning into the design process after you've made a Plan.
<u>APPRECIATIVE INQUIRY</u> Dream Discover Design Deliver	The vocabulary frees up our thinking and approach. Can cultivate lateral thinking and new angles. Originates from industry to assess practice and process.	No explicit analysis and evaluation steps could overlook important functions and elements in a design.

Summary of Analysis and Design Tools and Methods

TOOL / METHOD	WHAT WORKS WELL	WHAT IS CHALLENGING
PNI (positive, negative, interesting)	Quick solutions, great for comparing options and evaluating	Acronyms put some people off. Requires honest appraisal of all factors
SWOC (strengths, weaknesses, opportunities and constraints)	Useful insights, same as above, well known tool	Can feel mechanical and ends on a restrictive note
Thought Fest (formerly known as Brainstorming)	Frees up thinking without pressure of results. Especially good at the start of a design or when blocked	Puts off people who wish to focus on specific outcomes and info
Visioning	Opens the whole brain; uses senses, dreams and images to make providence happen	May be new or uncomfortable for some people
Observation: Listening to the Landscape	Captures creative and rational. Interesting to repeat over time	Imaginative mode always seems to yield the same observations
Mind Maps	Unlocks natural brain potential and organises info well	Best to learn some good techniques for optimum results
Pattern Thinking	Good when complexity and detail are overwhelming	Requires some familiarity to translate into designs
Key Planning Tools (sector, zone, elevation, network)	Vital for land use design. Accessible and logical	Takes time and lots of observation to map well
Input- Output Linkage	Enables clear connections and essential feedback loops	No guarantee all inputs and outputs will be included
Multiple Functions and Elements	Secures efficiency and resilience in the design	Wording can confuse some people
McCarg Exclusion Method	Simple, effective, protects from locating elements in the worst place	Relies on good knowledge of the site and people involved
Creating Guilds	Gives focus to beneficial relationships, creating an integrated system	Must remember to give attention to the connections between the guilds
Random Word technique	Unexpected and uncanny insights	Too random!
'Problem is the Solution'	Gives specific solutions; good for evaluation and tweaks	Requires some lateral thinking skill. Beware quick fixes

Repair Plants: Dynamic Accumulators

Here is some info and an activity on wild plants, commonly called weeds, more usefully termed 'soil repair plants'. These species are often maligned and misunderstood and yet they are vital for the well-being of ecosystems. They colonise every available space for a reason. As humans, we depend on ecosystems and so wild plants deserve our recognition and understanding.

Soil repair plants quickly cover any bare patch of ground to protect and enrich the soil. The plants below are known as dynamic accumulators, as it appears they actively accumulate minerals in their bodies, making them available for the soil and other plants. This process unlocks nutrient pathways, repairing soil structure and replenishing fertility, which in turn permits more complex interactions and larger species to return to the land, eventually becoming forest, our climax ecosystem.

Many of the repair plants are edible and/or medicinal; supporting a plethora of insects and subsequent processes of mitigating pests and disease; food for birds and frogs; pollination and more. By allowing areas for repair plants to flourish, we create a biodiversity and nutrient bank that benefits our place and beyond. We can also actively harvest these plants to boost our compost heap or to make liquid tonics to address mineral deficiencies in the soil.

SOME EXAMPLES OF COMMON PLANTS AND THE MINERALS THEY HARVEST

N — NITROGEN: alfalfa clover beans vetch comfrey kelp licorice root leaves nettles

K — POTASSIUM: bracken borage carrot leaves chamomile chickweed chicory clover dock comfrey eyebright fennel fat hen nettles mullein oak bark parsley peppermint plantains silverweed sow thistle watercress yarrow

P — PHOSPHORUS: barley buckwheat caraway chickweed clovers dandelion docks garlic fat hen lemon balm licorice root leaves lupine marigold meadowsweet mustards purslane savoy sorrel vetches watercress yarrow

Mg — MAGNESIUM: bladderwrack coltsfoot comfrey dandelion devils bit horsetails kelp meadowsweet mistletoe mullein parsley peppermint primrose sow thistle watercress willow bark

Mn — MANGANESE: chickweed eastern bracken fat hen

S — SULPHUR: coltsfoot eyebright fennel garlic meadowsweet mullein mustars nettles plantains shepherds purse watercress

I — IODINE: sasparilla bladderwrack devils bit dulse kelp

Cu — COPPER: coltsfoot eastern bracken nettles plantains silverweed sow thistle valerian yarrow

Co — COBALT: eastern bracken horsetails vetches

Si — SILICON: oat straw plantains valerian borage comfrey dandelion

YOUR LUCKY SEVEN

Select 7 plants from the list. Choose species you have heard of; would like to grow; or find to be growing at your place or nearby. See if you can cover most of the minerals listed with your Lucky 7. Encourage them in your wild edges, herb area or under fruit trees to be a backbone of biodiversity at your place.

(1)

(2)

(3)

(4)

(5)

(6)

(7)

NUTRIENTS IN SOIL AND THEIR FUNCTIONS

NITROGEN: essential for growth of leaves and stems
POTASSIUM: increases disease resistance and quality of fruits and grains
PHOSPHORUS: improves yields, promotes early maturity
CALCIUM: essential for cell division
SULPHUR: a constituent of amino acids, proteins and enzymes
IRON: helps in chlorophyll formation
MAGNESIUM: a major role in phosphorus transport
MOLYBDENUM: essential for nitrogen fixation
MANGANESE: assists chlorophyll production and metabolism depends on it
BORON: important in calcium uptake and quality of fruit and vegetables
COPPER: most functions are indirect and complexities
ZINC: helps in uptake and use of water in plants
CHLORINE: contributes to water holding capacity in plants

When facing the complexity of soil ecology, compost mixes and so on, a simple hack is to have a diverse mix of wild plants (weeds) growing around your more cultivated and tidy areas, that you regularly compost. Introducing varied Organic Matter (OM) frees up the natural potential and health in the soil, whether we understand the scientific complexities or not. To install this app in your brain, repeat OM (organic matter) a minimum of 7 times now.

Permaculture in Practice

Some ideas below from our experience...

OBSERVE AND INTERACT

"Beauty is in the eye of the beholder"

~

- Observed wind patterns to zone trees for windbreaks.
- Mapped sun angle over the seasons to find house and garden sites.
- Observed slug behaviour, made wider paths and planted more perennials to address this.
- Awareness of our energy levels informed our work patterns.

CATCH AND STORE ENERGY

"Make hay while the sun shines"

~

- Terraced the slope to collect water and nutrients in landscape.
- Grew comfrey to collect nutrients and store fertility in plants.
- Greenhouse and stone walls to store solar energy/ heat for plants and in our house.
- Planted lots of trees (solar collectors) for craft, fuel and food.
- Planted windbreaks to collect rainwater, protect crops and minimise soil erosion.
- Lazy gardening (allowed wild plants to grow all around) stored my energy.

OBTAIN A YIELD

"You can't work on an empty stomach"

~

- Grew potatoes to clear ground, cover soil and for food yield.
- Grass valued for sheep food and also scythed for nitrogen rich compost material.
- Planted nitrogen fixing trees and plants to enrich soil, cut for compost and for their fruit.
- Ducks gave eggs, slug control and the yield of fun company.
- Windbreaks improved fruit and veg yields.
- Trees gave huge biodiversity yield to many insect and bird species.
- Visitors and volunteers yielded feedback, help and positivity.

APPLY SELF REGULATION AND ACCEPT FEEDBACK

"The sins of the fathers are visited on the children unto the seventh generation"

~ Learnt perennial plants resilient to weather, pests and less work, so planted more.
~ Learnt through building chalet a better earth floor design and technique.
~ Introducing materials slowly allowed time to see their effects.
~ Tweaking garden plan every season to improve on mis-takes.
~ Moved barn after feedback from a local farmer, to a better site.
~ Sited trees in response to weather observations over 1- 5 years.

USE AND VALUE RENEWABLE RESOURCES AND SERVICES

"Let nature take its course"

~ Brash and willow provide simple craft solutions and items such as bean supports.
~ Natural building materials from the site e.g. stone, earth and timber.
~ Ducks as slug killers, avoiding manufactured or toxic products.
~ Solar design and water conservation in the buildings and land terraces.
~ Water and solar power provides us with electric.
~ Pigs to clear ground and provide yields of manure and potentially food.
~ Experimented with horse logging as an option to reduce our dependence on machinery.
~ Trees are a wealth of resources from fuel and soil repair to natural fence posts and food.

PRODUCE NO WASTE

"A stitch in time saves nine"

~ Compost everything that biodegrades.
~ Experimented with recycled plastic waste for insulation and drainage materials.
~ Reused old fencing wire from the fields for earth bag walls.
~ Brash from the firewood harvest highly valued as deadwood, rich in microbes.
~ Animal manures are valuable material for compost.

DESIGN FROM PATTERNS TO DETAILS

"Can't see the wood for the trees"

~

- ~ We used sector patterns (soil, air, sun, water, spirit), zone, aspect, elevation and networks to design layout both in the house and on the land.
- ~ The pattern of a woodland edge in the fruit gardens.
- ~ Siting the buildings and their elements (doors, windows etc) according to the patterns of sun angles, living use and air flows.
- ~ Edges for mineral rich plants and insect habitat.
- ~ Forest garden nursery inhabits a market edge.

INTEGRATE RATHER THAN SEGREGATE

"Many hands make light work"

~

- ~ Integrated all elements, e.g. ponds, structures, gardens, to get them in the 'right' place, mostly with success.
- ~ Ponds integrate the functions of collecting rainwater to irrigate the garden and greenhouse and as habitat for the ducks and other creatures.
- ~ Integrated trees through our designs for multiple functions, as outlined above.
- ~ Combining forces with willing helpers eased our work load and reciprocated with fun and information sharing.
- ~ Using the bath top as a potting table integrates the bath, dining space and greywater store for irrigation in the greenhouse.

USE SMALL AND SLOW SOLUTIONS

"The bigger they are, the harder they fall. Slow and steady wins the race"

~

- ~ Learnt to grow food gradually, didn't beat self up about mistakes.
- ~ Made a small house first to trial techniques for big house.
- ~ Implementation of the land designs has been gradual and staged to accommodate further observations on areas not needing immediate establishment.
- ~ Scything is slower than mowing, uses no petrol, is good meditative exercise and mashes up less frogs, giving them the chance to hop out of the way.
- ~ Mulching, used appropriately, not as instant as digging, yet improves soil health and is easier on the back.

USE AND VALUE DIVERSITY

"Don't put all your eggs in one basket"

~ Companion planting, flowers and veg, reduced pests e.g. aphids and slugs.

~ Our young forest garden is diverse in structure, niches, plants and habitat to create beneficial relationships and optimise yields.

~ Diverse mixes of trees wherever possible, such as 3 varieties of willow for insect habitat and general resilience.

~ Diversity of functions, e.g mixed nitrogen fixing coppice, wild edibles and fruiting hedge, as back up for fertility and food.

~ Dividing the field into diverse zones has made a bleak, under valued field into cosy and manageable areas.

~ Encouraging mycorrhizal fungi supports diverse networks between all our plants.

USE EDGES AND VALUE THE MARGINAL

"Don't think you are on the right track just because it's a well trodden path"

~ Whole smallholding marginal soil and on the edge of a hill, unique strengths e.g. view and brilliant opportunity to trial permaculture repair.

~ Dug ponds with shallow edges, ideal for frog spawn and other creatures.

~ Minimal digging preserved the top 1cm of soil, a most precious and alive edge.

~ Prioritising hedgerow and coppice edges, made wildlife corridor and habitat to join other existing woodland edges.

~ Demarking steep areas for grazing uses the edge effectively as sheep find it easier here than gardeners.

~ Using the marginal hedged corner in our west field for a barn provides weather shelter at lambing time.

CREATIVELY USE AND RESPOND TO CHANGE

"Vision is not seeing things as they are but as they will be"

~ Did our best to model ecological succession with nitrogen fixing trees and shrubs for shelter and perennials for food.

~ Areas are left to go wild to build soil health, habitat and for future resilience

~ The kitchen garden became a nursery for my small business, as new garden established.

~ Comfrey beds have been for initial soil building and then to be a resource for my livelihood (plant tonics).

~ Generally, our approach is a creative response to future challenges of extreme weather, and food and energy supply through an emphasis on resilience.

~ Paid attention to making the gardens and buildings suitable for us in our old age and for our children as they grew up.

Deep Time

Being aware of the challenges of our times, whether economic, ecological, societal or personal, we can be defeated by anxiety and overwhelm. Here follows a piece of writing I put together from some of my favourite authors that have helped me see with new eyes and go forth with empowerment.

With deep appreciation to Starhawk, Joanna Macy, Mary Jane Hodges and Eckhart Tolle, on whose work this writing is based.

A NEW MYTHOLOGY FOR AN ECOLOGICAL AGE

We live in a technological age, on a threshold of great change and possibly upheaval; millions of people are turning towards a vision of a new era, an Ecological Age. To help this transition, we can create a mythology for our times, harness the power of storytelling and visualisation to relieve us of our fears of inadequacy and pain and clear the way for empowerment, connection and creative action.

Joanna Macy talks of Deep Time, which means seeing ourselves in the bigger picture of the history of Planet Earth. We are increasingly faced with a world where there is so much more information to know and things and experiences to have. We are all faced with a lifetime and a world where we cannot know everything and cannot have everything. By having a sense of Deep Time we can find a sense of connectedness and unity amongst so many fragments. If the life of the earth is seen as a single 24 hour day, the history of humans is a mere minute before midnight. This is the story of Deep Time.

When we feel anger, anxiety or overwhelm about the world and events, we can transform that with awareness into will-power. It is not our own will that makes this transformation alone; something Deeper is pushing us all up the hill.

THE MAKING OF THE EARTH

A cold dawn, 4.5 billion years ago; clouds of gas and particles coalesced to form our planet. Our sun as yet too young to radiate heat and light, our earth, cold and dark. Constant meteors bombarded the planet.

THE MELTING POT

Over the next 2 billion years, half of our planet's life so far, earth's interior, under gravity and heat, formed a liquid iron core. Lighter materials created an outer layer of partially molten rock. The liquid surface belched hot gases of carbon dioxide, nitrogen and steam radiating out into space.

THE CRUSTED EARTH

Over the next 1.5 billion years the earth cooled to form its primitive crust, liquid rock, magma, surged to the surface as meteors hit the earth and rock spread out in vast sheets. Seething lava from enormous volcanoes solidified and the first continents floated on the rock mantle below.

FIRST SPARKS OF LIFE

As the gases billowed out from earth's hot interior for these billions of years, they cooled and condensed into clouds blanketing our planet. Massive thunder storms poured down for millenia and filled the young oceans. Violent electric storms made lightening that forged the elements into more complex molecules. The first streams and rivers appeared, eroding volcanic rock to form sand and mud on the ocean floors. The first organisms, basked in the warm waters of life. They feasted on the abundance of sunlight, floated in decadence in the oceans. They began to photosynthesise, trapping energy from the sun, transforming CO2 into carbon and oxygen.

THE AIR THAT WE BREATHE

By 2.5 billion years ago, the continents had begun to drift on their long wandering across the surface of the Earth. While the first breathers, the blue-green algae prospered, over millennia their activity began to replace carbon dioxide with oxygen. Some of that oxygen became ozone, screening the earth from the intense UV radiation of space and changed the composition of the atmosphere; creating the air that we breathe and making our surface waters safe for the emergence of more complex forms of life. Yet oxygen was deadly for those early life forms, wiping out 90% of their populations. A mass extinction created by their own activity. If we ourselves sometimes go to excess or fear scarcity and destruction from world excess, then take a breath and know it is a tendency of life to expand and contract.

THE FIRST ABUNDANCE

In the life of the earth as a day, at 9pm, 600million years ago, the foundations of earth as we know her today took form - a molten iron core; surrounded by a vast bulk of hot rock and the surface layer, as thin as eggshell in comparison to that which lays beneath us. Primitive jelly fish, soft corals and shelled animals began to appear. 200million years more, and plants began to move onto land; algae, mosses and ferns crept up rivers and estuaries in damp, tangled mats of green.

They were the pioneers.

These plants provided shelter, food, shade and most importantly moisture for animals to survive on land - giant millipedes and tiny spiders flourished amongst the tree ferns.

Then 100million years ago, life exploded again into diversity on land. Great tropical forests flourished and the earliest reptiles laid the first shelled eggs.

MASS DISAPPEARANCE

A jostling and a drifting of the world's land masses and Pangea had emerged, forming many mountain ranges we would recognise today. These shifts in continents and polar ice melts led to half of all species alive at this great entrance of life to disappear, three quarters of all amphibians and reptiles became extinct and the giant forests covering much of the earth died away.

RETURN TO LIFE

It took a mere 20million years for earth to rebound to biological diversity this time, just 3 minutes in earth's long day so far, half an hour or so before midnight. Many new species appeared; first snakes and early dinosaurs. At night, our first true mammal, a tiny shrew, crept about unseen.

The planet, now 2/3 water and ice; our atmosphere a thin veil of vapour. The rivers of the world, transporting billions of cubic metres of water a day to the oceans. Our atmosphere, the gases on which life depends, clutched to earth's surface by gravity, without which earth would be baked by day and frozen by night.

Blue skies, rainbows, sunsets that blaze like fire, are all effects of this precious atmosphere.

THE EMERGENCE OF HUMANITY

Some 10's of millions of years ago (or Deep Time minutes) many animal families again disappeared. Thirty seconds before midnight and the first evidence of early humans arrive. Birds now took to the air, seed bearing plants with huge flowers now conquered the world and mammals began to flourish.

DEEP TIME

Like a mountain stream, the flow of life seems to have gone through long quiet periods, interrupted by moments of sudden, dramatic change. At each of these great turning points, half or more of all living organisms, both in oceans and on land have quite rapidly disappeared. Those species that survived, emerged into a world of opportunity, rich with ecological niches left empty by the death of their predecessors. At these moments, organisms have suddenly radiated into a proliferation of new forms.

Just as plants lived for millions of years and then suddenly evolved to flower and rocks witnessed millennia before metamorphosing into seams of crystal: we can know, that a similar potential for transformation is possible for us.

For 3 million years humans have wandered the earth, hunting and gathering and for a mere 10,000 years, earth has witnessed the first settlements, such as the Olmec in Mexico and Machu Pichu in Peru. In ancient Babylon, complex laws, urban structures and religions were born; and over millennia human wanderings across the earth have been epic and adventurous. From Stone Age Polynesian voyages through to the middle ages when Europeans set off across the globe with their flocks and crops, pests, plants and cultural traditions.

So now we are perhaps at another crossroads in history - species extinction, extreme weather events, the seasons in flux and polar ice caps seemingly in transition. The scale of human suffering inflicted on a species by its own kind is beyond our comprehension. Yet more than ever we are connected, have access to awareness and the power to make conscious choices at such a Great Turning.

First, we can imagine our ideal world, weave the story of it. Breathe the reality of a New Earth into being:

What values would you dream to see treasured in society; who would our monuments in parks and plazas be dedicated to; who would be our heroes?

What kind of homes might we have? Picture the textures and materials we could use. Can you imagine new ways to provide a home with heat, water, air and light. In your ideal world, what would you wish for in a home and in the design of your neighbourhood?

How would we care for our elderly folk and educate our children? Perhaps you can hear a future where voices of all individuals and groups are integrated and diversity of perspective enriches harmony between people?

You might dream of new forms of agriculture and trade; or evolution in the way work is organised and the way we govern. Perhaps you can sense another way to fund healthcare, education and environmental protection rather than from the profits of production and consumption.

What would an earth look like with peace, equality and respect for All Beings?